AMAZING ANIMALS

AMAZING ANIMALS

By the Editors of Time-Life Books

TIME-LIFE BOOKS, ALEXANDRIA, VIRGINIA

CONTENTS

1
SHAPES FOR SURVIVAL

Diversity that enriches life and makes it possible: The paradoxical platypus . . . deep divers . . . birds that swim . . . walking fish

2
FIGHTING FOR LIFE

Weapons of attack and defense: An ancient arms race . . . stuffing and puffing . . . animal actors . . . cooperative couples

3
OVING AND LIVING

Mating, nesting, and rearing the young: Babysitting in Antarctica . . . invitation to the bower . . . long-term love . . . midwives . . . murderous interlopers . . . beaver builders

4
TRAVELERS

Master travelers of the animal kingdom: Butterflies and milkweed . . . the caribou highway . . . a never-ending flight . . . a crab feast . . . urban invaders . . . Huberta the hippo

5
THE ANIMAL MIND

Intelligence and emotion among animals: A genius monkey . . . a "talking" chimp . . . grieving gorillas . . . elephants' silent speech . . . animals under the influence . . . kind-hearted vampires

SHAPES FOR SURVIVAL

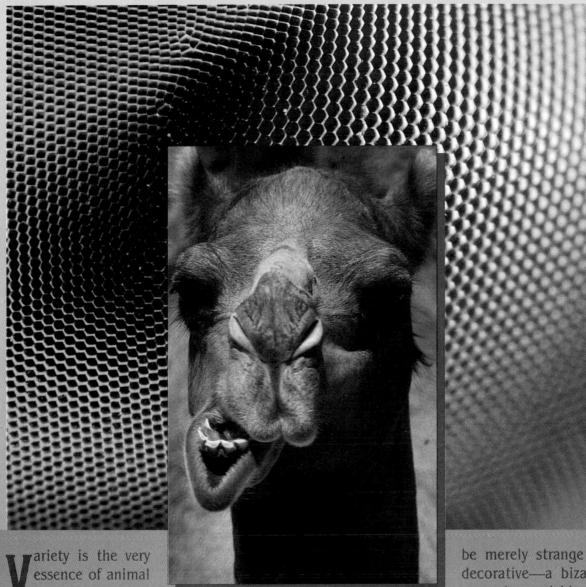

Variety is the very essence of animal life. The world is full of flying, crawling, and burrowing creatures. Feathered, furred, and naked, fat and lean, huge and tiny, animals exhibit every shape and behavior. Birds fly, but some can only walk or swim. Fish swim, but some climb onto land. Predators thrive by virtue of their keen eyesight, but some are blind. Yet the variety has a single purpose: What appears to be merely strange or decorative—a bizarre appendage, vivid color, or unusual marking—is necessary for survival. The variety is endless, and new discoveries are made every year; although scientists are often baffled by these findings, they do know that not a feather, fiber, bone, or sinew exists that does not, or did not at one time, further life. The diversity of evolution not only enriches life—it makes it possible.

The Platypus Paradox

In 1798, puzzled scholars of London's British Museum were confronted with the pelt of a platypus from the faraway penal colony of Australia. Their reaction was predictable, if unscientific: The worthy sages assumed that this unlikely object was an example of irreverent Australian humor and began cutting it up to determine how it had been assembled from miscellaneous animal parts.

The foot-long pelt still hangs in the museum, the marks of the suspicious scholars' scalpels still visible where they tried in vain to find the stitches that would prove that this odd beast was nothing but a practical joke. The British Museum's skeptical view of the platypus is certainly understandable, but if the animal is a joke, the joke has been played by nature, not by human pranksters.

The platypus is several species rolled into one, an egg-laying, aquatic mammal with physical characteristics borrowed in equal measure from mammals, reptiles, and birds. The unique qualities are more than skin-deep: Genetic researchers have found that the platypus has chromosomes typical of both mammals and reptiles and shared by no other creature.

Its long, leathery bill and large webbed feet seem to be borrowed from a duck; its broad tail and thick coat of waterproof fur could come from a beaver. Birdlike, the female has only one reproductive ovary, but like a lizard she grows the eggs in her genital tract until conditions are ripe for hatching. After breaking out of their shells, the baby platypuses act like mammals, drawing nourishment from their mother's milk.

The male defends his territory and his mate by wielding hollow, inch-long spurs on his ankles, capable of injecting an opponent with a venom powerful enough to kill a dog or cause severe pain, swelling, and partial paralysis in a human being.

The platypus is considered to be an ancient creature—fossils of similar animals date back 100 million years—but not a simple one. Scientists continue to discover new attributes of the platypus, even after 200 years of intensive study.

It was not until 1980, for example, that an explanation was found for the animal's ability to remain underwater for as long as ten minutes without breathing. The secret lies in an unusually large number of red blood cells, which enable the platypus to retain a greater amount of oxygen than most other mammals can.

In 1986, surprised researchers announced another startling discovery: The platypus has what can only be described as a sixth sense. When the animal is swimming underwater, flaps of skin tightly seal its eyes, nose, and ears, leaving it temporarily deaf, blind, and unable to smell. For many years scientists assumed that, thus encumbered, the animal used an acute sense of touch to hunt for food.

But investigation revealed that the platypus relies instead on a far more precise detection system, homing in on its various prey by sensing the faint aura of electric fields they create. Freshwater shrimp, a platypus favorite, generate weak electric charges with each flick of their tails. As the platypus hunts, it swings its rubbery bill from side to side, allowing some 850,000 sensors to detect and locate the shrimp's slight electric emissions with uncanny accuracy.

The sensitivity to electricity al-

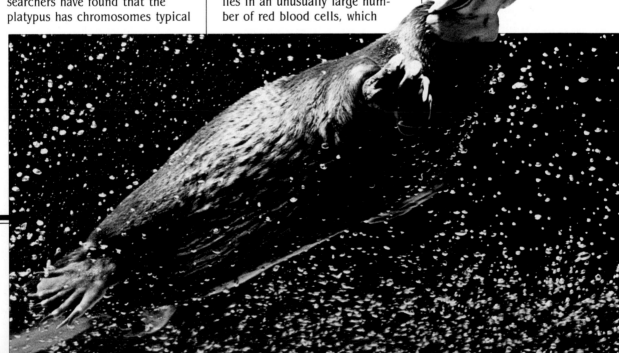

Excellent insulation enables penguins
such as this emperor, shown diving
in frigid antarctic waters for squid, to
rank among the most completely
aquatic of living birds.

so may help the platypus navigate underwater by detecting barely perceptible electric fields generated by water flowing over obstructions, such as rocks or logs. □

On Ice

Exposed to the bitter cold winters of Alaska and Canada, the North American wood frog freezes almost solid. Two-thirds of the water in its body turns to ice; if one of its extremities is twisted, it snaps off. The frog's heart stops; if cut, the animal will not bleed. But, despite all the appearances of death, a seemingly frozen wood frog is very much alive. With the advent of spring warmth, the frog will resume its normal activities.

When most animals freeze, they die because ice crystals rupture their body cells. But when the wood frog is exposed to freezing cold, its body responds by draining water from cells, then surrounding them with a freshly produced supply of the natural sugar glucose. This substance serves as an antifreeze, preventing ice from forming in the remaining water and saving the frog's internal organs.

The frog is not alone in its ability to freeze and thaw. The Arctic carabid beetle, for example, uses glycerol, a sweet, syrupy substance derived from fat, to remain active in temperatures as low as five degrees below zero Fahrenheit. Even at temperatures as low as sixty below, the beetle's glycerol, like the wood frog's glucose, protects it from ice damage. When the weather turns warm, the beetle returns to a normal, active life. □

Antarctic Air Conditioning

The emperor penguin's body is so well insulated that the bird sometimes must find a way to cool off, even in the frigid antarctic winter.

The penguin's familiar black and white coloration ornaments a three-layered insulating system that begins inside the skin with a thick blanket of blubber. What little heat escapes the fat is trapped close to the body by a zone of air chambers formed between tufts of light down growing from the shafts of the bird's feathers. The penguin's outer defense is a tough layer of curved, overlapping feathers, made waterproof by oil. Thus protected, the emperor can plunge freely into the freezing waters of its homeland and spend hours diving after the fish and crustaceans that make up its diet.

On land, however, exertion or a slight rise in air temperature can cause the emperor to overheat. But its structure contains a radiation system that deals with heat as effectively as with cold. Blood vessels lining the blubber layer expand, increasing the circulation of heated blood near the skin, where it is cooled by outside air. The flightless penguin speeds the cooling process by flapping its wings and ruffling its outer feathers to increase air circulation. □

Design for Diving

The Weddell seal, which spends most of its life swimming beneath the ice of Antarctica, can dive to a depth of 2,000 feet in five minutes, travel more than a mile in a single dive, and remain submerged for up to an hour. The Weddell seal accomplishes these feats with ease, thanks to a body that is engineered to withstand the rigors of the life it leads.

Internal structures protect the seal from the bends—which afflict humans during deep dives, when air breathed under great pressure is absorbed by body tissues. If a diver ascends too quickly, air bubbles form as the pressure is released, causing pain, paralysis, and sometimes death.

To prevent the bends, human divers must ascend slowly, allowing the air bubbles to pass away harmlessly. But the Weddell seal's body imposes no such limitations. Just before diving, the seal exhales, reducing the volume of air in its lungs and other tissues. As the pressure builds during the descent, the animal's flexible rib cage collapses, pushing still more air out of the lungs and into rigid, impermeable airways from which it cannot pass into the blood stream. As a result, the Weddell seal can dive deeply and shoot to the surface as quickly as it likes without any ill effects.

The diving seal's oxygen supply is contained in a great reservoir of blood that is half again as large as the blood supply of other mammals of similar size. While submerged, the Weddell seal conserves oxygen by slowing its heartbeat to one-tenth its normal rate and shutting off circulation to muscles not needed for swimming, directing it instead toward the all-important heart and brain.

When the Weddell seal finally needs to come up for air, it swims to one of several breathing holes it has bored in the ice. If the bitterly cold antarctic air has formed a layer of ice over the hole during the seal's absence, the beast readily cuts its way to the surface with forward-pointed upper front teeth that appear to be designed just for that purpose. □

The Antarctic's common Weddell seal is a highly efficient diving machine.

Solar Bear

The polar bear carries its own solar heater—a coat of transparent, hollow hairs that serve as light pipes, which some scientists say are capable of transmitting 95 percent of the sun's invisible, warmth-giving ultraviolet rays to the bear's skin.

Although its hair is transparent, the polar bear appears white or yellowish, because its coat reflects most of the visible light striking it. But ultraviolet rays are trapped within the hair walls and conducted down to the skin—which is really black—to be absorbed as heat.

The same hairy coat and a thick layer of fat beneath the polar bear's skin trap body heat so effectively that the animal is virtually invisible to heat-detecting infrared sensors. □

THE EARTHWORM senses light in much the same way that humans feel warmth. Light-sensitive cells in its skin allow the worm to differentiate between light and dark—a sense sufficient to keep the worm safely underground and away from deadly predators and the drying rays of the sun.

Fleas in Flight

A flea does not really jump; it is launched, accelerating 50 times as fast as NASA's space shuttle rocket and soaring 150 times its own length, the equivalent of a 6-foot human leaping 900 feet.

Two elastic pads, made of a substance called resilin, power the flea's takeoff. To prepare for flight, the flea kneels and compresses the pads, much as a jack-in-the-box cover compresses the coil spring hidden inside. Then it locks its legs in place with two hooklike appendages that allow the flea to remain cocked and ready to leap without further muscular effort. When the hooks are released, the resilin pads spring back to their original shape, flinging the flea upward in tumbling flight *(below)* until its six bristle-covered legs, which are tipped with sharp hooks, grab at its intended host.

If at first the flea misses its target, it tries again, and again, and again; fleas are virtually tireless. Naturalist Miriam Rothschild—one of the world's foremost experts on fleas—recorded one taking off 30,000 times without pause. □

The Right Stuff

Hummingbirds are the acrobats of the avian world. A typical intricate aerial ballet begins when a bird hurls itself from its perch, reaching top speed almost instantly. Then, twisting and turning, flying forward and backward, hovering one moment and racing sixty miles per hour the next, it flashes from flower to flower and dips its needlelike beak deep to drink the nectar. For a show-stopping finale, the hummer may spread its tail, execute a quick backflip, and fly upside down.

The hummer's flying prowess is derived from the strength and design of its wings. Powered by flight muscles that account for nearly one-third of the bird's body weight, hummingbird wings are a blur of motion—78 beats per second in normal flight and up to 200

during one of its wild courtship dives. The wings' rapid motion creates the humming sound from which the bird takes its name.

Flexible shoulder joints allow hummingbird wings to flap in a figure-eight pattern that can be instantly altered to suit the needs of flight. Rigid wing tips assure that every movement has an immediate effect on the air.

To support its frenzied activity, the hummer's heart beats 1,260 times per minute, and in proportion to the bird's size it is larger than that of any other bird. Even at rest, the hummingbird breathes 250 times a minute. The hummer burns energy so rapidly it must take in almost two-thirds of its body weight in flower nectar each day, the equivalent of a 150-pound man eating 90 pounds of food. □

High-speed strobe photography captured this Anna's hummer hovering and feeding at a gooseberry branch.

This swift and nimble housefly, shown below dining on placemat crumbs, may carry a population of over two million bacteria on its body.

Stunt Pilot

The common housefly, whose eating habits and reputation as a disease carrier make it the target of relentless attempts at extermination, has developed uncanny powers to protect itself from the fly swatter and other hazards.

The fly's maneuverability in flight—at speeds that can exceed fifty miles per hour—surpasses that of the most nimble jet fighter. The shape and construction of its wings and body keep the insect flying no matter what its speed or what its position.

The insect's flight is stabilized with the help of two vibrating, knobbed stalks extending from its body; sensors at their bases detect tilting motions. The fly's wings have no muscles of their own; instead, two sets of flight muscles are connected to a flexible thorax, the central part of the fly's body, whose vibrations can flap the wings at more than 20,000 beats per minute. The flight muscles work like opposing rubber bands; when one contracts the other stretches, each energizing the other through several cycles before running out of energy. A strong starter muscle sets them in motion; the cycle of alternate stretching and contracting continues for 20 strokes or so before another pull by the starter muscle is needed.

Acute vision and delicate sensors in the forms of antennae and hairs give the fly a lightning-fast reaction speed of less than one two-hundredth of a second—ten times quicker than that of a human hand.

Thus able to dodge predation, the fly is free to travel and feed. Hairs on its feet are coated with an

adhesive substance, allowing the fly to walk upside down on ceilings or across slippery glass—and to transport all manner of decaying debris and harmful bacteria. Fly feet contain sensors that enable the insect to taste-test potential meals and therefore avoid those containing pesticides. Most flies, however, happily dine on meals that are laced with DDT; they have developed a resistance to this once-common chemical poison.

For all their protective mechanisms, large numbers of houseflies do succumb to the exterminator and the swatter. Yet the insect survives and prospers due to its stunning ability to reproduce. One mature fly, laying its first eggs in the month of April, could claim 191,010,000,000,000,000,000 direct descendants by August, if all were to live and breed. □

Diminutive Hitchhikers

The tiny, wingless rove beetle is the tropical rodent's home exterminator, trading its services for food and transportation. The beetle is carried from nest to nest by mice and rats in the forests of Central and South America, so that it can purge their dwellings of mites and fleas.

The tiny bug hitchhikes by clinging to the hair on a rodent's neck and head; the grateful carrier, perhaps anticipating respite from the torments of its parasites, pays no notice, even when the beetle runs across its eyes and whiskers.

Other tiny hitchhikers prefer to travel by air. One species of mite lives its life inside blooming flowers, eating their nectar and pollen. But when the food supply declines in one flower, the mite flies to a new home by scampering up the long beak of a feeding hummingbird and sheltering in its nostrils. There it rides until the bird stops to feed at a likely looking prospect for habitation.

The wingless bee louse attaches itself to the body of a honeybee, stealing its food from the bee's mouth. At times, great numbers of the lice will pick the same bee and ride it right into the hive, where they enjoy a plentiful feast. □

Parasitic mites get a lift to a new source of food on the back of a daddy-longlegs.

Hot Music

Among the characteristic sounds of summer is the incessant chirping of the male American tree cricket. Rubbing a serrated vein on one wing against a scraper on another, the insect broadcasts a pervasive love song that females of the species appear to find irresistible.

Because its tempo increases as temperatures rise, the cricket's call can serve as a thermometer: To determine the temperature, just count the number of chirps in ten seconds, then add thirty-nine. □

Efficient Fireflies

The humble firefly, winking coyly at the children who pursue it on a summer evening, is one of nature's most efficient creatures, converting 95 percent of its available chemical energy into light.

The firefly's yellow-green glow is produced by mixing the chemical luciferin, an enzyme called luciferase, and oxygen in the bug's light organs on its abdomen. The process gives off a cold, efficient light; only 5 percent of the energy expended is wasted as heat. In contrast, a light bulb wastes 90 percent of the electrical energy flowing through it.

The firefly's glimmer is used to attract a mate, usually without any fear of attracting unwelcome predators; an internal toxin makes the bug distasteful to most insect-eating animals. □

Huge, compound eyes with almost 30,000 facets—each a complete eye in itself—makes dragonfly vision among the best in the insect world.

Bug-Eyed

Two creatures looking at the same thing often see totally different pictures. Dragonflies, for example, do not see details visible to humans. But more important to the dragonfly's survival, its compound eyes afford a wider field of vision and the ability to detect movement forty feet away.

Other visual specialties abound among animals. The flying fox, a bat that relies on sight rather than hearing, has malleable lenses and variable focusing. The result: built-in telephoto vision. And the chameleon, with eyes that operate independently, can look in two directions at the same time.

Contrary to popular myth, most animals do not see only in black and white. But color perception varies. Were a human to see with a bee's eyes, red roses would appear black and bright purple flowers would dot the landscape. And the bee's five multilensed eyes are sensitive to the ultraviolet hues common to many flowers, which act like brilliant beacons marking the way to pollen. □

Hearing the Light

The darting, acrobatic flight of bats swarming from darkened daytime resting places to hunt their insect prey in the night appears to be a spastic ballet danced in silence. For, although most bats are blind and hunt by sound—a skill called echolocation—human ears can hear only the rustle of their leathery wings.

To a bat, however, the night air is filled with high-pitched shrieks and screams that rend the air like sirens. As they fly, bats emit clicking noises at ultrasonic frequencies far beyond the range of human hearing; the returning echoes locate and identify obstacles, other bats, and the insects that constitute most bats' food. The system is so sensitive that it can locate a flying insect or an obstacle barely thicker than the edge of a dime.

The clicks become more frequent as a bat closes on a quarry. At a distance, the area is broadly scanned at about 10 clicks per second; when prey is "sighted," it is identified with a rush of sound at 25 to 50 clicks per second. As the bat moves in for the kill and follows the evasions and maneuvers of its victim (such as the moth above), the sounds rush forth as fast as 200 clicks per second.

To a bat, each click sounds more like a siren's whoop, undulating from high pitch to low to high; as the clicks' speed increases, so does the range of the whooping frequencies. The volume of bats' signals is like a siren's, too—or as one researcher described the sounds, "like the screams of a jetliner at close range."

At the same time that a bat must protect itself from such piercing blasts, its hearing also must be sensitive enough to detect the slightest returning echo. To accomplish this, the bat literally switches its hearing on and off with each click. Middle-ear muscles that transmit sound vibrations from the eardrum to the inner ear reduce the volume by contracting when the bat emits its resounding clicks. Then they quickly relax to detect the faint echo.

The bat's echolocation skills are similar to the complex, computer-aided device known as sonar, by which submarines navigate and fishermen locate their quarry. In fact, the development of sonar in the late 1930s led to the confirmation of the bat's long-suspected navigational expertise, proving that the animal is not, as was once thought, guided by some mysterious sixth sense. □

Dinner by Candlelight

The eyesight of the night-prowling owl was long thought to be insignificant next to its remarkable hearing. In fact, the great nocturnal hunter probably relies even more on its eyes than on its ears. An owl's sight is 100 times as acute as that of a human, enabling the bird to spot the movement of a mouse by candlelight 100 yards distant and pick an insect out of the air in near-total darkness.

While human eyes—and those of most other animals—are spherical, those of the owl are tubular, efficiently piping light to the retina. This sensor at the rear of the eye is densely packed with the light-sensitive pigment rhodopsin, also known as visual purple. Consequently, more light reaches the owl's retina, where it is more effectively sensed than in most other animals. Because the eyes are so large, there is no room for muscles to move them. To compensate for this lack, the owl has fourteen vertebrae in its neck—human beings have seven—giving it the disconcerting ability to swivel its head through three-quarters of a complete circle.

Like humans but unlike most birds, the owl has binocular vision, providing it with the accurate depth perception required for efficient hunting.

The owl's famous ears are almost as formidable as its eyes, gathering the slightest sounds and pinpointing their source accurately. Vertical flaps of skin extending from the top of the owl's head to its jaw capture sound and funnel it into the ears. The shape and location of the ears are such that each one hears the same sound somewhat differently—louder or softer, arriving sooner or later. The owl is thus able to locate the source of the noise by comparing the differing perceptions.

For all its remarkable eye and ear structures, however, the owl's real sensory magic may be accomplished within its brain. According to scientists, this complex organ is uniquely prepared to process the abundant information gathered by the bird's eyes and ears, literally providing an accurate mental map of the owl's surroundings and potential prey. □

The eyes of the great horned owl are so large that there is no room in the bird's head for muscles to move them.

Providing a large surface area, the ears of the desert fennec fox help dissipate heat.

Providing a large surface area, the ears of the desert fennec fox help dissipate heat.

Ear Conditioning

When an African elephant spreads its ears in the wind and sprays them with water shot from its trunk, it is turning on its highly effective air-conditioning system. With a surface area of as much as thirty square feet apiece and thousands of blood vessels laced throughout, an elephant's ears amount to vast cooling coils that can lower the temperature of blood flowing through them by as much as sixteen degrees Fahrenheit.

The elephant shares this attribute with several smaller species. The fennec fox, whose ears poke six inches above its head; the jug-eared jerboa, a European rodent; and the North American jack rabbit all count on their ears for cooling. The jack rabbit can also lower its ears over its back, providing a living blanket on cool nights. □

Suntan Lotion

The reddish secretions that ooze from the skin of the hippopotamus, which once led people to believe that the animal sweats blood under the tropical sun, are instead a natural sunscreen. Curious researchers have found that the pink, viscous substance works as well on human beings as it does on hippopotamuses. □

Falling Cats

One fine spring day in 1984, a cat named Sabrina plummeted thirty-two stories to the sidewalk from an open window of her New York City apartment. The result no doubt enhanced the belief that cats have nine lives: Sabrina lived, suffering only a chipped tooth and a partially collapsed lung.

Although remarkable, Sabrina's survival was by no means unique. Cats do have an uncanny ability to survive falls—and according to scientists, Sabrina the cat was more likely to survive her 320-foot plunge than one of less than 100 feet.

Air resistance and the cat's acrobatic ability to position itself in midair help it survive such deadly drops. Although in theory all objects fall at the same speed, in real life they do not. The friction of the air balances the inexorable pull of gravity, so that every object descending through the atmosphere—cat or cannon ball—has a maximum speed, called its terminal velocity. Heavy, streamlined objects, such as cannon balls, fall a long way before reaching a very high maximum speed. Lighter things that offer resistance to the air reach terminal velocity quick-

ly, and then fall much more slowly.

A cat's speed peaks at about sixty miles per hour after the animal has fallen 100 feet—the height of a ten-story building. Until it reaches this terminal velocity, a falling cat seems just as disoriented as any other animal, and for that reason, cats falling fewer than ten stories are quite likely to be injured or killed.

However, once the animal's speed stabilizes during a longer fall, a cat quickly prepares for landing. It relaxes and flips upright, spreads its legs to increase air resistance, and flattens its body to spread the force of impact. The cat lands with its legs bent, further absorbing the shock. □

In this stop-action photography sequence, a cat instinctively rights itself during a fall.

Two Left Feet

The South American hoatzin, which looks like a chicken with a bright blue hairdo, is born with four "feet," and the young birds swim and climb trees long before they learn to fly.

The hoatzin chick first leaves its nest three days after hatching and immediately starts climbing trees by grasping overhead branches with the footlike claws on its wings. The unkempt nest is built over water, so that the chicks can dive and swim away from any possible danger. When the threat has passed, the birds use their feet to climb back up into the nest.

The chick learns to fly at the age of six weeks, and it soon sheds its extra claw-feet. With them goes much of the hoatzin's mobility: An adult hoatzin flies clumsily, walks with some difficulty, and can stand only by leaning on its breast, which is armored with a thick layer of skin. □

Unique footlike claws on its wings enable the hoatzin chick to cling to branches and follow behind its parents and its siblings, begging to be fed.

The Gentle Touch

The star-nosed mole, which lives most of its life in total darkness, may be the most touch sensitive of all animals. Twenty-two petal-like, semiretractable pimples encircle its bright pink snout like tentacles, allowing the blind mole to feel its way through its burrow and to sense the movements of its favorite meal, the earthworm. □

Animal Myths

The chameleon is one of nature's great frauds, taking credit for a trick it cannot perform: deliberately changing color to match its surroundings. Its stand-in, the tiny green anole lizard—which is often misrepresented as a chameleon—cannot achieve such color changes either. Although both reptiles alter their hues in response to danger, heat, and light, any similarity to their background, scientists say, is probably coincidental.

The chameleon myth is only one of many that surround members of the animal kingdom. The hyena, for example, has long suffered from its reputation as a cowardly scavenger that slinks after the majestic lion in hopes of scrounging the remains of a kill. Not so,

says a Dutch zoologist, Hans Kruuk, whose observations in Tanzania in 1967 revealed a highly capable hunter that kills two-thirds of its own food. In fact, says Kruuk, it is the lion who is the more frequent scavenger.

Another misconception has it that ostriches hide their heads in the sand. In fact, when faced with danger, the ungainly bird sits down, stretches its long neck, and rests its head flat on the ground. This posture is supposed to fool predators into thinking that the flightless ostrich is just another grassy mound.

Other animal myths have equally shaky foundations. The black panther is a nonexistent species, although there are anomalous black

leopards with still-blacker spots. Bats do not try to fly into women's hair, even though a disoriented one might do so by accident. And penguins are not confined to frigid Antarctica: Indeed, one species lives as far north as the Galápagos Islands, just below the equator. □

An excited West African chameleon, shown here with offspring, might display bright green coloring after a fight, though brown could offer better camouflage.

Wet and Dry

The camel's legendary ability to thrive in hot, dry environments has made it an indispensable beast of burden throughout North Africa, Arabia, and central Asia.

Pliny, the first-century Roman naturalist, maintained that camels accomplished their desert survival feats by storing large quantities of water in their stomachs. Others have claimed that the animal's hump (or two humps, in the case of the Asian camel) serves as a vast reservoir.

Neither is correct. A camel's hump is actually a mound of fat—stored energy—rather than water. The animal's affinity for desert regions does not stem from storage but from feats of water conservation made possible by a physique that is more wondrous than Pliny ever imagined.

A camel's nose is lined with absorbent tissue that captures and recycles two-thirds of the moisture it exhales. Even stray liquid leaking from the nostrils is trapped in nasal channels that funnel it through the camel's cleft upper lip and into its mouth.

The camel's water conservation is assisted by its tolerance for wide fluctuations in body temperature, from a nighttime low of 93 degrees Fahrenheit to a daytime high of 105 degrees. Only at that high temperature does the animal begin to sweat, giving up some of its precious water supply to cool its body. Even then, the camel can afford to lose more water than most other animals can—as much as 30 percent of its body weight, compared to only 12 percent for human beings.

Having increased the odds of survival in the desert through water conservation, the camel guarantees its value as a worker with great powers of recuperation; a weak, emaciated, and dehydrated camel can put its body right with a single ten-minute drink of about twenty gallons of water. When the supply is available, a camel can consume more than thirty gallons of it in a day.

Other desert dwellers have developed equally effective mechanisms for water conservation. The kangaroo rat, for example, has no sweat glands and produces urine so concentrated it solidifies soon after being excreted. And some small creatures, such as the Australian koala—whose name comes from an aboriginal word meaning "no water"—exceed even the camel and kangaroo rat in their tolerance for dryness: They never need to drink at all, drawing all the moisture they need from their food. □

The camel's nose *(left)* is uniquely suited to aid the animal in moisture retention. Even more efficient at retaining moisture, the desert-dwelling kangaroo rat *(above)* rarely consumes liquids, while Australia's koala *(right)* never drinks water at all, ingesting nothing but eucalyptus leaves.

Great Size, Complex Structures

Like a living skyscraper, the giraffe looks down upon nearly all other living things. On land, only the elephant compares in size, its one and a half ton bulk as dominant on the landscape as the giraffe's two-story height. Although size alone makes both elephant and giraffe notable, there is also much to admire in the extraordinary details of their great bodies.

Scanning the landscape as it browses in the tops of thorny acacia trees, the giraffe has little difficulty seeing and avoiding its few natural enemies. The animal's long legs can carry it away from trouble with an effortless thirty-two-mile-per-hour lope; if flight fails, a threatened giraffe can decapitate a full-grown lion with a kick of its twelve-inch hooves. Even human hunters—the most deadly of predators—have been foiled by the giraffe's tough hide, which is so thick in places that it resists soft-nosed bullets fired at close range.

Internally, the giraffe is equipped with a uniquely engineered circulatory system to maintain a constant flow of life-giving blood—no small matter when the animal stoops to drink and abruptly lowers its head nearly seven feet below its heart. Without some mechanism to regulate its circulation, this could be fatal, causing a cerebral hemorrhage as blood flooded into the lowered brain. The animal would also be in constant danger of fainting when it raised its head, due to blood rushing out of the brain.

To maintain a constant flow of blood to its brain regardless of how quickly it moves its head, the giraffe has a network of reservoir-like blood vessels and one-way check valves just below the brain.

Sloths descend from their arboreal lofts
less than once a week—to defecate.
Unable to walk upright, they must drag
themselves along the ground.

Its heart muscle is enormous and powerful, with three-inch-thick walls pumping twenty to twenty-five gallons of blood under the highest pressure of any mammal—twice that of the average healthy human. To withstand the pressures and carry such a volume, the giraffe's veins and arteries are as much as an inch in diameter, with thick, elastic walls.

Like the giraffe, the elephant has a uniquely endowed body. Unlike the giraffe's, the elephant's neck is so short and muscular it cannot bring its mouth to food and water. Thus, its seven-foot-long trunk—an extension of the upper lip and nose—serves as a hand to transport food and water to the elephant's mouth. In times of anger, the trunk becomes a devastating weapon. Forty thousand muscles and tendons make the trunk strong enough to lift a large log, yet so sensitive that it can pluck a single blade of grass.

The elephant's tusks, fearsome weapons against the beast's few enemies, are also important food-gathering tools. Growing as long as eleven feet, tusks are used to splinter trees to bare their soft inner pulp and to gently pry yams from the earth.

The elephant grinds its food—some 400 pounds of forage every day—with a pair of giant molars. Six sets of teeth grow in an elephant's jaws, one below the other. As each set is worn down by the beast's fibrous diet, another emerges to take its place. But when the last tooth is gone, the elephant cannot eat and will starve to death in the wild. □

Hanging Out

The sloth, that sluggish tree dweller whose very name is synonymous with laziness, lives in the rain forests of Central and South America, clinging to tree limbs with its long, meathooklike claws. It eats, sleeps, mates, gives birth, and nurses its young upside down, and has so adapted to this topsy-turvy life that it can turn its head 270 degrees, permitting it to view the world right-side up while its body hangs upside down.

The sloth's reputation for lethargy is a well-earned one. The animal sleeps fifteen to eighteen hours a day; even when roused, its movements verge on the impercep-tible. So still is the sloth that algae grow in its hair, providing the animal with a rough green camouflage that blends well with its leafy environment.

From millennia of inverted tree dwelling, the sloth has virtually lost its ability to move on the ground. It is barely able to stand upright and cannot walk at all; it must drag itself on its belly. On one occasion, it took a sloth forty-eight days to travel four miles. But most sloths avoid any form of locomotion: Some are said to live their entire lives in one tree, and they often remain hanging even after death. □

Submarine Birds

The common loon, whose ghostly cry haunts the lakes of North America, can reach flight speeds of more than seventy-five miles per hour. But the loon spends most of its time in the water *(above)*, where it dives skillfully—sometimes remaining submerged for as long as two minutes—and swims faster than most fish.

Twisting and rolling its mottled black body, a swimming loon could be mistaken for a miniature seal. Its legs tuck within its body to maintain a streamlined form. Even its bones are adapted to an underwater life: They are solid, not honeycombed with air sacs, as are the light, buoyant bones of other species of birds.

The ocean-going gannet spends most of its time in the air, but it, too, has adapted to its watery world. The gannet spends hours each day cruising above the waves in search of fish; when a meal swims into view, the bird plummets from the sky at sixty miles per hour, plunging as much as fifteen feet below the surface to grab its quarry.

To cushion the shock of the high-speed impact with the sea, the gannet has air sacs surrounding its skull and some extraordinarily strong bones in its wings. Both nose and eyes are protected from seawater; valves close within the nostrils, and transparent membranes cover the eyes. □

Grounded

The ostrich, which may stand eight feet tall and weigh more than 300 pounds, is the largest bird in the world. Taking twenty-eight-foot strides, it can run as fast as a racehorse and easily outdistance most predators *(below)*. Those who somehow catch it must deal with a kick hard enough to bend half-inch steel bars and toenails sharp as a knife—with which it swiftly disembowels its foes. But for all the ostrich's abilities, there is one thing it cannot do: Like a handful of the world's birds, it is unable to fly.

Although the ostrich has wings, it lacks the bone and muscle to make them work.

The emu, a 100-pound flightless Australian cousin of the ostrich, shares the ostrich's speed. In an incident dubbed the Emu War, the homely bird's forty-mile-per-hour sprints humbled a crack Australian army unit charged with eradicating the thousands of emus that had taken to foraging on the croplands of southwestern Australia. At the end of the month-long campaign in 1932, the soldiers had managed to kill only twelve of the birds.

Not all flightless birds can count on speed for survival. The chicken-size kiwi, New Zealand's national symbol, is clumsy afoot and spends its days hiding in underground burrows. When it comes out at night, it is ideally equipped to ferret out food. With nostrils located at the end of its beak (all other birds have them at the base), the kiwi can smell a worm three inches below the surface.

The flightless ostrich, emu, and kiwi are descendants of birds that did fly millions of years ago. Scientists speculate that a lack of natural predators rendered flight unnecessary, and, over time, the birds' flight muscles atrophied and the heavy bone structure needed to support the muscles disappeared.

Not so the penguin. The stubby wings of this bird do wave ineffectually in the air. But its skeleton and flight muscles are well developed, and it uses its flipper-wings to swim at almost twenty miles per hour. Feeding on fish and small crustaceans and bobbing up for breath every few seconds, penguins sometimes swim continuously for three to five months while migrating, traveling as much as 900 miles and pausing only occasionally to rest on a handy ice floe. □

Walking Fish

A fish out of water is usually a dead fish—unless it happens to be a mudskipper, which is common in Africa and the islands of the Pacific Ocean. This peculiar creature is so much at home on land that it climbs on fallen trees and flees toward land, not water, when it is threatened.

The gill cavities of the mudskipper are lined with folds of skin richly laced with blood vessels, enabling the fish to absorb oxygen from both air and water. It takes its name from its movement on tidal mudflats, where it forages for worms, insects, and mollusks with a hopping, skipping motion, propelled by strong crutchlike pectoral fins and a powerful tail. It uses the fins to grasp tree bark tightly and pull itself upward; other fins act as suction cups, allowing the skipper to climb smooth surfaces.

Another part-time land dweller and tree scaler is the climbing perch that inhabits the waters of Southeast Asia. It has divided gills that allow it to breath air and remain out of water for days. □

These mudskippers are so at home on land that they court in shallow water *(right)* and readily climb trees *(below)* looking for food.

Flying without Wings

The flying dragon, or butterfly lizard, of Southeast Asia is a spectacular sight as it soars through the rain forest on brilliant orange and black wings. Like a number of other aerial animals, however, this six-inch-long lizard is not a true flyer. Rather, it is a glider, holding itself aloft with loose flaps of skin that it can stretch to create the effect of fixed wings.

The four-foot-long giant flying squirrel of Southeast Asia uses a similar technique, which enables it to glide as far as one-quarter of a mile, thus bridging the ravines of its mountainous territory.

The North American flying squirrel cannot glide as far as its Asian cousin, but it is an uncommonly nimble aerial acrobat. Twisting and turning its body and stretching membranes connecting front and hind limbs, the squirrel executes midair U-turns, lateral loops, and spiral descents that make it an elusive quarry for its enemies.

The colugo, a flying lemur in Southeast Asia, is equally agile in the air, thanks to a cloaklike membrane extending from neck to tail that it controls by spreading its arms and legs. Leaping from a lofty perch, the colugo often swoops 200 feet in a single glide, maneuvering by twisting its body from side to side and using its tail for balance.

Even some snakes can "fly." The East Indian paradise tree snake, for example, coils in trees awaiting its prey. When it springs toward its target, the snake spreads its ribs, flattening its body into an aerodynamic shape that enables it to glide as far as 100 feet, literally snaking through the air to make a perfect strike. □

This Alaskan flying squirrel can easily change flight direction during a fifty-yard glide by simply raising or lowering its legs. It uses its tail as a stabilizer.

Long-Term Survival Rations

Modern-day crocodiles have shrunk from the fifty-foot beasts that prowled the swamps of 100 million years ago, but they remain the largest living reptiles. Saltwater varieties average fifteen feet in length, and one fearsome specimen measured twenty-eight feet.

The crocodile's survival is largely due to its undiscriminating and voracious appetite. A crocodile can fast for as long as six months, then devour an entire horse over the course of two days. A full-grown crocodile fears no wild animal: It will tackle ferocious Cape buffalo, rhinoceros, and elephants. Even sharks that stray into shallow estuaries are fair game.

One hungry crocodile in Australia once took a chunk out of the propeller of a fifty-horsepower outboard motor as it roared past. In the stomach of another croco-dile—which was killed after it had devoured a grown man—hunters discovered a four-gallon metal can and two blankets.

Crocodiles routinely carry in their stomachs five to ten pounds of stones, which help grind up their food and serve as ballast, offsetting the buoyancy of their lungs so they can lurk half-submerged until an unsuspecting meal comes by *(below)*. □

This dozing ground squirrel will spend the greater part of its life hibernating.

Eating It All

The common cockroach has eaten its way through 300 million years of the earth's lifetime, outliving scores of other species—among them the dinosaurs—to become one of the oldest surviving forms of life on earth.

Roaches will eat virtually anything; they thrive on such diverse delicacies as shoe polish, soap, and paint. They relish fingernail clippings and will chew the eyebrows off a sleeping human. If pressed, a roach will devour its cast-off skin, its own eggs, or other roaches. One variety of the insect is able to digest wood, like a termite, and when times are really tough, some roaches can survive three months without food and a month without water.

As delectable as roaches find the world, the world finds them repugnant. Most predators turn up their noses at the noxious fluids and odors exuded by squashed roaches. The cockroach's survival is further enhanced by its near indestructibility. Swat one and its hard shell may save it; remove its head and it will live on for several days. Subject it to desert heat, polar cold, or nuclear radiation, and the cockroach will thrive.

The roach's ultimate survival secret: No matter what the conditions, cockroaches continue to breed. A single pair can produce 400,000 offspring in a year. □

Deeper than Sleep

The best-known animal response to cold is hibernation, a sleeplike state—but deeper and longer lasting than sleep   during which many complex changes take place to keep the creature alive. The hibernator's temperature drops and breathing slows, sometimes to the point that it is almost imperceptible. Metabolism is drastically reduced, so that the animal can live off stored body fat.

But the Arctic ground squirrel goes hibernation one better—it allows itself to cool below freezing, to twenty-seven degrees Fahrenheit. Only every three weeks does the squirrel arise from its winter sleep, warming to its normal ninety-nine degrees and leaving its burrow to eliminate wastes. Baffled scientists have found no natural antifreeze in the ground squirrel's blood or any other mechanism that allows it to chill without freezing.

The North American black bear, on the other hand, adopts a state of near hibernation, sleeping intermittently throughout the seven-month winter break. Female bears rouse themselves long enough to give birth to one to three cubs, which remain in the den with their sleeping mother, nursing until she wakes in the spring, when they all sally forth into the world. □

Popcorn provides a gourmet meal for a hungry cockroach.

31

Sleepy Time

For many years, the mournful nocturnal song of the Nuttall's poorwill—a relative of the whippoorwill—had entertained Edmund C. Jaeger on his summer camping trips in the Chuckwalla Mountains of southern California. But never in winter; from October to March, the bird and its song disappeared. Jaeger, a biology teacher at California's Riverside College, joined other scientists in assuming that the bird, like many others, migrated to a warmer territory.

But the assumption was dashed and ornithological history made one cold December morning in 1946. As Jaeger and two companions explored the Chuckwallas' jumbled, rugged canyons, they found a healthy adult poorwill nestled snugly in a crevice.

The creature neither flew away nor trembled with fear when Jaeger touched it. In fact, the bird seemed totally unaware of his presence. Light directed into its eyes drew no response, a mirror held to its nostrils did not fog, and a stethoscope could not detect a heartbeat. Yet the bird was alive, and eventually it sluggishly opened its eyes and stretched its wings.

The real reason for the poorwill's winter silence was now apparent: This species, alone among all birds, hibernates. Further examination revealed that this was true hibernation, not the occasional torpidity commonly employed by birds to protect themselves from cold weather. The poorwill's temperature, at 64.4 degrees Fahrenheit, was far below its normal 106 degrees, and its respiration was almost at a standstill.

Jaeger returned to the canyon each winter for the next four years, each time finding the same bird— by now identified by a metal band—nestled in its crevice, nearly invisible because its mottled feathers closely matched the frozen granite surroundings. One winter the poorwill's hibernation lasted eighty-eight days.

Ironically, Jaeger later learned that, although science was in the dark, the Indians living among the Chuckwallas knew of the poorwill's hibernation all along. On one of his visits to observe his history-making bird, Jaeger asked a Navajo child where the poorwills went in winter. Without hesitation, the boy replied, "Up in the rocks." □

FIGHTING FOR LIFE

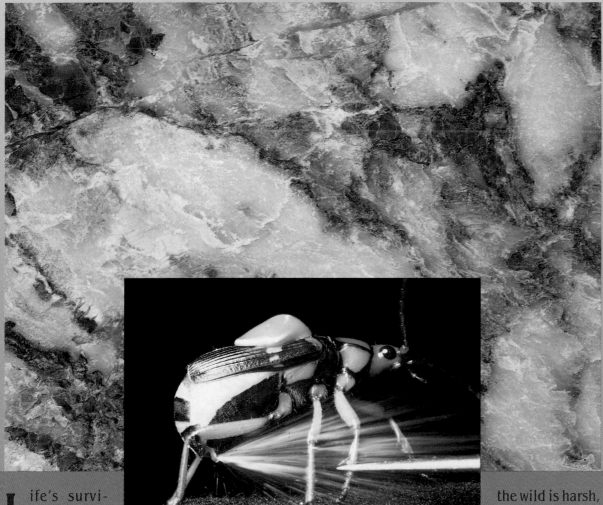

Life's survivors are outfitted with a great number of weapons and skills to assure a steady supply of food—and to avoid becoming food for others. Fang, claw, pincer, and barb are the weapons of attack and defense, employed not only by the mighty, but by the humble as well. In the contest for survival, the greatest of beasts are the least vulnerable, but the smallest warriors of the animal world are among the fiercest, possessed by appetites that rage insatiably within their bodies. But although life in the wild is harsh, not all feeding involves savagery; and not all defense requires a lethal response. Clever deception and cheerful cooperation produce food and ensure safety as often as a fearsome charge. Quiet, artful camouflage and steely nerved bluff often defend as effectively as tooth and claw. And animal actors as resourceful and clever as any human win their livelihood by artifice and invention, making the world an arena in which they play an all-or-nothing game of life.

Marauding Masses

Army ants of the Western Hemisphere and their African cousins, the so-called drivers, are tropical migrants as fearsome in reality as in reputation. Like armies on the move, these minuscule marauders—as many as 20 million at a time—swarm over crops, homes, domestic animals, and wildlife, taking what they wish and leaving behind a ravaged land.

The driver ants of Africa are the more destructive, having sharp mandibles well suited to cutting and tearing the flesh of vertebrate victims as well as vegetation and other insects. Anticipating the ants' arrival, terrorized animals take flight. Those too small or feeble to escape are killed and carried off to be eaten. Injured or tethered goats, horses, and dogs have had the flesh stripped from their bones.

Although folklore no doubt exaggerates the prowess of ants, the 8,800 known species are creatures to reckon with. Weighed on a scale, ants would account for 10 percent of the mass of the world's land animals. And their versatility in devising ways to feed and protect themselves is considerable.

Some varieties battle and enslave other ants, becoming so dependent on slave labor that they no longer know how to dig their own nests or feed themselves or raise their young. Other ants flourish by engaging in peaceful agriculture rather than war. Dairying ants, for example, capture and domesticate aphids in order to milk them regularly for the sweet droplets of honeydew the aphids excrete. In the Tuxtla Mountains of southern Mexico, carpenter ants form a similar relationship with butterfly caterpillars, actually transporting the grubs to food sources in exchange for their honeydew—a risky business for the ants, since the mature butterflies feed on ant larvae. And *Hypoclinea* ants are nomads, driving herds of sap-feeding mealybugs to richer pastures so the ants can feed on nutrients in the bugs' excrement.

Leaf-cutting ants include some 100 varieties of small, spike-covered farmers that cultivate miniature gardens of fungus on pieces of leaves, which they chew and then store in underground compost piles. Their colonies often harbor several million ants, and

Dairying ants milk honeydew from aphids *(left)*. Weavers are known their intricate nests *(above)*.

the garden chambers can extend as deep as twelve feet underground.

Many ants are foragers and savers: In the Middle East, such ants gather the few seeds that are to be found in the arid places where they dwell and store those seeds in underground granaries. In North America's southwest desert, honey-storer ants deposit reserves of sweet plant secretions in the abdomen of one of their own, which becomes a bloated and immobile emergency fuel tank.

Some ants do appear to sacrifice themselves for the good of the colony. Certain species living in the desert of North Africa repeatedly forge into the world beyond their underground nests, returning again and again with food until disappearing, hopelessly lost or eaten by spiders and flies. During their brief lives of unceasing labor, they bring home up to twenty times their own weight in food.

Many ants hunt their prey, but others are trappers too. Among these is the *Dacetini*, found worldwide, whose quarry includes that most elusive of insects, the springtail—which, as its name suggests, is able to launch itself out of the way of most predators. The *Dacetini*, however, is armed with a pair of long, sharp mandibles that it sets like a trap. It lies on the forest floor, awaiting a springtail's jump to release the hair trigger.

Even the most domestic breeds of ant can be aggressive. Weaver ants, natives of the forests of Africa, Southeast Asia, and Australia, assemble colonies of silken nests that they defend against all intruders, including humans, with bites of their mandibles and sprays of blistering formic acid. □

Arms Race

There are few events in the rich and varied insect world of tropical Africa more noted for their fierce drama than the swift, deadly search-and-destroy missions periodically launched by *Megaponera* ants on nests of termites.

Led by a lone scout, these large black ants, specialists in termite warfare, march in columns to the termites' nest. There, they swarm every hole and crevice of the victims' stronghold and, one by one, drag the termites to the surface. The raid concluded, the columns of warriors form once again, each ant carrying three or four maimed termites. As if proclaiming their triumph, the ants make shrill screeching sounds as they march back to their own nest.

In fact, the *Megaponera* raiders are but one species of warrior ant that has taken part in an evolutionary arms race between ants and termites that has been extended through most of the 100 million years of their existence. Attackers—usually the voracious, omnivorous ants—have developed a variety of weapons with which to assault the steadfastly vegetarian termites, who are usually thrust into the role of defender.

For example, the Malaysian Basicerotine ants are built for squeezing into narrow spaces, and their mouthpieces are perfectly formed clamps with which they grasp fleeing termites. The jaws of other species of ant soldiers are so adapted for fighting that the ants are unable to pass food to their own mouths, depending instead on the services of smaller laboring ants for nourishment.

Although they are seldom aggressors, the termites are ardent defenders. Several varieties of termite soldier are adept at chemical warfare, hitting their attackers with noxious secretions. Another termite is described by scientists as an "anti-ant atomizer"—it can accurately shoot a jet of sticky liquid from a glandular gun, halting the attackers in their gummy tracks. But the tactic sometimes backfires: In the heat of battle, the termite sometimes pumps so violently that its body explodes, sliming everything within range. □

Gluttonous Goats

In 1773, during one of his voyages to the South Seas, Captain James Cook found his ship burdened with a surplus of goats, which were carried to provide fresh meat for the ship's crew. His solution was to unload the unneeded animals on the shores of New Zealand's two main islands. Years later, when the lush paradise was colonized, more goats were sent to provide meat for road workers and miners and to eat away the runaway growth of bramble and bracken.

In addition to their assigned tasks, however, these prolific and voracious animals stripped trees of bark, devoured shrubs, glutted themselves on ground mosses, and cleared much of the plant life that protects topsoil from erosion. They chewed some native plants to the edge of extinction.

The goat's reputation for unbridled appetite is not exaggerated.

Weighing up to 260 pounds, it shares with sheep a taste for straw and grass. But the bearded goat can easily stomach paper, wood, and garbage. Its intestines harbor microbes that break down hard-to-digest cellulose, making even linen hanging from a clothesline tempting fare. Poisonous yews that normally kill horses and cows merely give this iron-bellied beast indigestion. Worse, its surefootedness enables the goat to defoliate rugged mountainsides and even climb trees to strip their limbs.

With the exception of humans, probably no creature bears more blame for environmental catastrophe. The bleakness of much of the Mediterranean landscape, where bare rock tops the hills and only goat-proof scrub brush survives on lower slopes, is attributed not only to heat and arid winds, but to centuries of grazing by goats. □

A Diet of Worms

The North American wood turtle, unable to sing for its supper, has learned to dance for dinner. With a gentle rocking motion, the turtle stamps its feet, causing vibrations in the ground that drive earthworms to the surface, where the turtle plucks them up.

Researchers speculate that the worms may mistake the vibrations for the patter of raindrops and squirm upward to avoid drowning. Another theory is that the worms are fleeing what they believe to be the approach of a hungry mole. Whatever the reason, other animals are wise to the trick: European gulls and plovers and the flightless kiwis of New Zealand all have been reported to stomp up worms.

Turning the tables, some animals gather their meals by sensing rather than creating vibrations. A spider lurks at the edge of its web waiting for the slight tremor that occurs when an unlucky insect has become trapped. And the African clawed frog senses the movements of tiny insects on the surface of the water so acutely that it is able to strike its victims accurately even in the dark. □

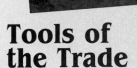

Left to right: A sea otter floats a rock on its belly to crack clams; a chimp uses a stick to dig for termites; an Egyptian vulture prepares to drop a stone on an egg.

Tools of the Trade

A story is told that the great Greek dramatist Aeschylus met an untimely end in 456 BC when a vulture dropped a rock on his prominent bald head. The poet's pate, the story goes, was mistaken for a delectable ostrich egg, which the rock was intended to crack.

The story's historical accuracy may be suspect, but not its description of rudimentary tool use by animals. Sea otters, for example, dine while floating on their backs, cracking the abalone they relish on large stones balanced on their bellies. Sticks and similar objects are also favored as hunting and feeding tools. The woodpecker finch of the Galápagos Islands and the green jay of Texas use cactus spines and sharp twigs to spear insects. Chimpanzees prepare special twigs for poking into termite nests, carefully withdrawing the insects that adhere to the twigs and licking them off. Chimps have also been known to clean their teeth with sticks and use leaves as toilet paper. They have even fashioned primitive sponges from chewed leaves to collect rainwater from deep tree holes.

Perhaps the most macabre tool in the animal world is used by several species of shrikes and Australian butcherbirds. They catch insects, grasshoppers, lizards, frogs, and smaller birds, which they impale on thorns and barbed wire for storage. □

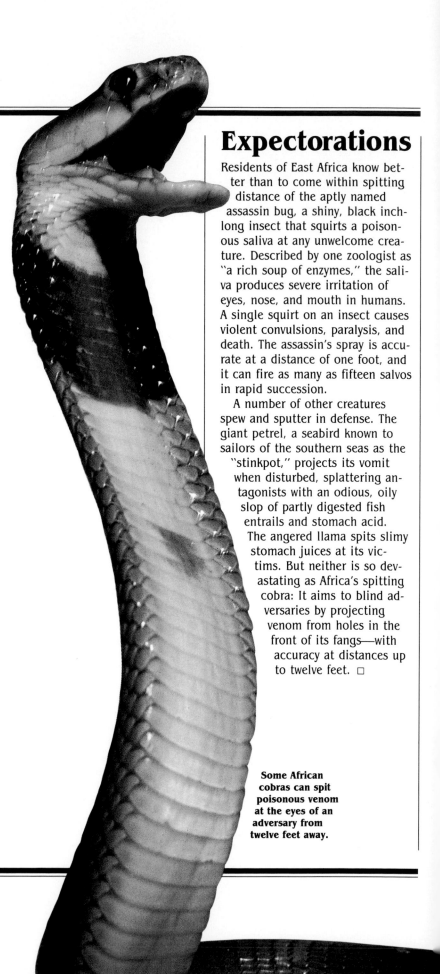

Stuff, Puff, and Protect

Because its tasty meat is a favorite food of predators, the foot-long chuckwalla lizard often gets itself into tight squeezes. In fact, that is its best defense.

When threatened by an attacker, the broad, flat chuckwalla scrambles for a crevice in the rocky outcrops characteristic of its habitat in the southwestern United States. Once inside, the lizard gulps air, puffs itself up, and becomes so tightly wedged that most animals cannot dislodge it. However, the region's Indians long ago learned simply to burst the chuckwalla's balloon with a knife or a stick.

Other creatures, such as the earthworm and lugworm, grip their burrows so firmly and persistently that they are torn apart before letting go. But the African pancake tortoise may be the chuckwalla's only companion in "puffery." Its soft, flexible shell can collapse almost flat, enabling the tortoise to slip into a tight place before inflating itself securely. □

Expectorations

Residents of East Africa know better than to come within spitting distance of the aptly named assassin bug, a shiny, black inch-long insect that squirts a poisonous saliva at any unwelcome creature. Described by one zoologist as "a rich soup of enzymes," the saliva produces severe irritation of eyes, nose, and mouth in humans. A single squirt on an insect causes violent convulsions, paralysis, and death. The assassin's spray is accurate at a distance of one foot, and it can fire as many as fifteen salvos in rapid succession.

A number of other creatures spew and sputter in defense. The giant petrel, a seabird known to sailors of the southern seas as the "stinkpot," projects its vomit when disturbed, splattering antagonists with an odious, oily slop of partly digested fish entrails and stomach acid. The angered llama spits slimy stomach juices at its victims. But neither is so devastating as Africa's spitting cobra: It aims to blind adversaries by projecting venom from holes in the front of its fangs—with accuracy at distances up to twelve feet. □

Some African cobras can spit poisonous venom at the eyes of an adversary from twelve feet away.

Boom Bug

Despite its diminutive size and attractive black and yellow coloration, the bombardier beetle should be considered armed and dangerous. Equipped with what amounts to a turreted cannon at its rear end, this ground beetle can fire a thirty-round chemical barrage that stinks, stings, and startles. Fired at a rate of six per minute, the beetle's blasts so confuse its foes that the bug often escapes unscathed, if not unnoticed.

When threatened, the half-inch-long beetle mixes the chemicals hydrogen peroxide and hydroquinone (the same chemical that stink bugs emit) with enzyme igniters, in a special hardened chamber. The combination releases oxygen, which provides the propulsion to fire corrosive quinones.

With each audible pop, a stream of hot quinones is discharged from a gland at the rear of the beetle, vaporizing into a foul-smelling tear gas powerful enough to disable insects, repel frogs, and irritate human skin. □

A round of "bombs" discharged by the bombardier beetle may distract an attacker, allowing the bug to escape.

Playing Possum

Although it is considered one of the stupidest and least aggressive of animals, the Virginia opossum must be doing something right: It is a survivor from the days of the dinosaurs, and it is the last of the pouched marsupials to be found in North America.

The size of a fat house cat, the opossum, though slow afoot, is an adept climber. But rather than scamper up a tree when threatened, the opossum simply plays dead. Dropping to the ground in a limp heap, it strikes a catatonic pose. Its eyes half shut and glassy, its mouth agape, teeth exposed, tongue dangling, and emitting a greenish ooze from an anal gland, the creature remains motionless for as long as six hours. An attacker can poke it, bite it, bark at it, and pick it up, but the opossum will not move. More often than not, predators—preferring to eat only what they kill—lose interest, because they assume the animal has already died.

Although "playing possum" has earned the animal a place in American idiom, the trick has not won much credit from scientists. Most think the opossum simply lacks the intelligence to devise the act deliberately. Rather, they speculate that the creature faints from fear, its nervous system overloaded to the point of temporary paralysis. □

Needling the Enemy

In 1952, the state of Vermont paid a fifty-cent bounty for each pair of porcupine ears, hoping that hunters would be encouraged to thin the state's burgeoning population of the prickly rodents. But twelve months and $90,000 later, there were still too many of the animals prowling the woods, supplementing their ordinary diet of nuts, berries, and bark with such tempting man-made delicacies as plywood, canoe paddles, tool handles, and automobile tires. The bristle pig, as this shy herbivore is sometimes known, had again proved its superior skill as a survivor.

It is hardly surprising. The porcupine has few natural enemies. It wishes to be left alone and makes a keen point of it. With a coat covered by as many as 30,000 quills bearing needle-sharp black tips, an angry porcupine is a daunting opponent. Tightening its skin so the spines spread and stand at attention and rattling them like swords in battle, the ten- to twelve-pound pincushion backs into its enemy, swinging its tail with enough force to embed hundreds of quills in any creature unfortunate or foolish enough to stand its ground.

Removing the hollow quills is an agonizing process. Body heat and moisture cause them to expand, driving thousands of microscopic barbs into the victim like fishhooks, firmly anchoring each shaft and assuring that flesh will be removed along with the quill. There is no alternative; left in place, porcupine quills can prove fatal. Muscle movement pulls each quill deeper and deeper—as much as an inch a day—causing infection and injury to internal organs.

Quills discovered in polar bears, deer, owls, and dogs attest to the fact that many animals tangle with porcupines. But few do so as a matter of course. The fisher, a fierce weasel-like animal, is known for its skill at flipping a porcupine on its back and tearing open its soft underbelly, but even fishers have been found blinded or choked to death by quills. (Vermont finally got a grip on its porcupine problem by introducing fishers into the porcupine-infested areas.)

Human predators, who sometimes take a fancy to roast porcupine—reputed to be something of a delicacy—have occasionally bitten off more than they can chew. In one instance, surgeons operating on a man complaining of severe stomach pains found that a quill had skewered his intestine. It had been ingested with a carelessly prepared porcupine sandwich four days earlier. The man died. □

The three-banded South American armadillo can roll into a ball and completely close itself off from attackers.

Hardball

The shell-like bands that cover the armadillo appear to make this animal a shoo-in for survival. But it is not invulnerable. Even the three-banded armadillo of South America, which can curl its body into an armored ball and roll away from danger, remains at risk from large carnivores whose wide jaws can crack its shell.

There are twenty varieties of armadillo, all of which deserve their name—which in Spanish means "little armed one." Throughout 55 million years of existence, these timid, musky-smelling mammals have found a number of ways to supplement the protection offered by their medieval-looking armor. Some scurry into thorny thickets, where their shells protect them from the spines that stop hunting dogs and other animals in their tracks. When cornered, the six-inch-long fairy armadillo of Argentina and Bolivia burrows into the

ground and plugs the opening with its plated rear end.

Not all defensive actions are successful. The unfortunate nine-banded armadillo leaps straight up when startled, a show of aggression designed to frighten away its enemy. But the armadillo's most dangerous modern enemy is the automobile, and highways are littered with the bodies of animals that have bounced into the undersides of speeding cars. □

Animal Oscars

In the animal kingdom, acting skills can often make the difference between life and death. And the burrowing owl may be one of nature's best performers. A ground-nesting bird about nine inches tall, this inhabitant of grasslands from the United States to South America is preyed upon by coyotes, badgers, cats, and snakes. But when cornered inside its burrow, the owl mimics the rattle and hiss of a rattlesnake, often discouraging predators.

Less convincing are the theatrics of the killdeer and grouse hen, which feign injury to their wings in order to distract predators from their chicks. The act works best with a young, inexperienced audience: Adult dogs and foxes—perhaps once denied a meal by their own gullibility—learn to recognize the act for what it is and search the area all the more intensely.

Costumes appear to be as important to animal actors as to humans. Through the first months of life, defenseless infant cheetahs bear the colors of the adult ratel, a particularly ferocious badger that is avoided by most larger animals, even hyenas and packs of dogs. Only later, when a cheetah is becoming dangerous in its own right, does its coat change to a distinctive black-spotted tan.

Although mimicry is most often employed for defense, at least one creature—an aggressive species of firefly—uses it to satisfy its appetite. The female *photuris* firefly displays the signals that females of several other firefly species use to attract mates. When the duped suitors arrive, she devours them. □

Small Terrors

The smallest of all North American mammals is also one of the fiercest: The three-inch-long pygmy shrew weighs only one-eighth of an ounce, but it is driven by a roaring metabolic furnace that must be fueled ceaselessly. The tiny terror consumes its weight in food every three hours, killing and devouring animals twice its size in seconds, without pausing to distinguish between flesh, fur, and bones. All portions are necessary, for the shrew will starve if it is deprived of nourishment for a day. Although the pygmy is the smallest North American shrew—the two-inch fat-tailed shrew of Europe is the world's tiniest mammal—all varieties share its appetite and ferocity. For most, including the pygmy, sharp teeth suffice as weapons. But the short-tailed shrew paralyzes its victims with a venom, chemically similar to that of the cobra, excreted from its salivary glands.

The shrew's habits illustrate a truism of nature, that the smallest creatures are not necessarily the weakest. The ratel, a twenty-five-pound, ten-inch-high African badger that stinks like a skunk, is armed with powerful claws, razor-sharp teeth, and a savage disposi-

A water shrew displays its ferocity as it seizes upon a comparatively large frog, which the shrew will devour completely—bones and all.

tion that see it through most encounters. Further defense is provided by a nearly impenetrable hide that hangs over its squat frame so loosely that the badger can twist within its own skin, enabling it to strike back even when grasped firmly in an enemy's jaws. So endowed, ratels have killed huge buffaloes and driven away packs of wild dogs.

In North America, the wolverine is considered so heedless of the normal order that to some Indian tribes it is the devil incarnate. Unpredictably vicious, more cunning than a fox, and far more powerful than its forty-inch, thirty-pound size implies, the wolverine is a member of the weasel family, although it looks more like a small brown bear. The animal is too slow to catch most prey, so its meals

result from bullying other animals away from their own kills. Mountain lions, grizzly bears, and even packs of wolves are known to back

away from a hungry wolverine. And there are reports that the animals have torn through the timbers of a cabin to get at the food inside. □

Disposable Parts

When a denizen of the tropical forest wishes to dine, it frequently selects the gecko lizard for an entrée. Sometimes, however, the diner must settle for an unpleasant hors d'oeuvre—the gecko's distasteful, detachable tail.

Threatened by a predator, the gecko has the unusual ability to break off its own tail by twitching a muscle and snapping a vertebra. No

blood is shed, as vessels in the gecko's body quickly seal. Meanwhile, nerves in the tail make it twitch, distracting the attacker and allowing the now-tailless lizard to escape from danger.

Salamanders, tadpoles, and a number of other lizards also have the option to abandon parts of their bodies, usually tails, rather than serve as an enemy's lunch or dinner. And, like the gecko, they grow the parts back.

Although an abandoned tail sometimes makes a satisfactory, if unpleasant, meal for a predator, the lizard itself usually will not. The gecko's tail simply tastes bad, but its other lizard parts contain toxins that are potent enough to kill a larger animal. □

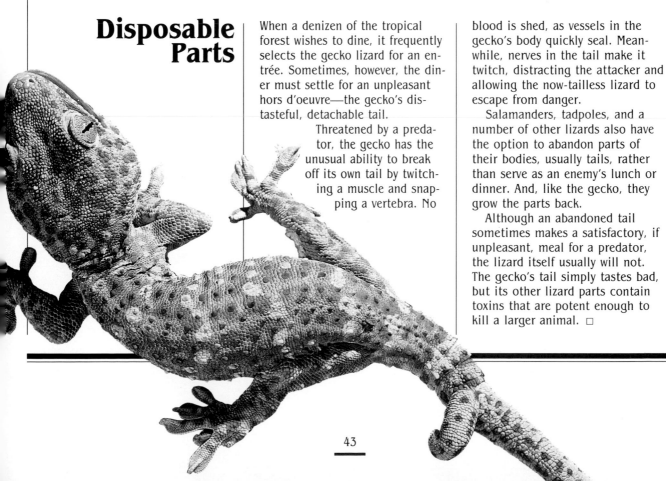

The warm-weather plumage of the willow ptarmigan renders the bird almost invisible in its summer surroundings.

The warm-weather plumage of the willow ptarmigan renders the bird almost invisible in its summer surroundings.

Only the dark eyes and nose distinguish this Eurasian ermine from its white winter habitat.

Now You See It . . .

The Alaskan ptarmigan's seasonal change of plumage from summer's moss brown to winter's snow white makes it one among many of nature's wallflowers. For creatures such as the ptarmigan—a variety of grouse—and the snowshoe hare, to stand out from their surroundings spells certain doom.

But others can play at this game of deceit. Sharing the same tundra with the ptarmigan and the hare are their predatory enemies—the arctic fox, the short-tailed weasel, and the snowy owl—all of which change colors with the seasons so they can stalk their prey by blending into the landscape.

Sociability rather than costume is the guise preferred by the North American zone-tailed hawk. This predator, much feared by small animals, keeps company with vultures, which the ground dwellers consider harmless because the large birds prefer carrion to live prey. Thus hidden among the vultures, the hawk watches for unsuspecting victims on which it pounces without warning. □

The arrangement and shape of these treehoppers make them look like inanimate thorns.

Hiding Out

Preparing for a meal, the ambushing inchworm, an unlikely predator found in the Hawaiian Islands, bites a piece out of a leaf's edge just big enough for itself. Tucking its body into the gap and withdrawing deadly claws, the inchworm thus concealed looks to all the world like nothing more than a harmless bump on a leaf. To a fly approaching too closely, the deception is invariably fatal; true to its name, the worm springs from ambush and makes its kill.

Inchworms take their name both from their size and peculiar gait, by which they appear to measure the world inch by inch. They are masters of disguise, plastering themselves to twigs, blending with leaves, and even chewing off pieces of flower petals, which they stick to their backs in order to conceal themselves on blossoms.

But inchworms only do what seems to come naturally to many of the earth's 30 million insect species, some of which turn the forests and fields into a deadly serious game of trick or treat. Indian stick insects imitate brown grass stalks, waving in the wind exactly like their models. An inchlong, leaf-eating weevil in New Guinea carries a garden of fungi, algae, lichens, and other plants on its back, approximating its surroundings to the extent that it tolerates tiny mites and other parasites in the flora. Some assassin bug nymphs plaster their backs with bits of termite nests, so they can inconspicuously lie in wait by the door of their prey. Planthopper larvae in East Africa deceive en masse, clustering together to imitate flowers on a stem.

Birds are sickened by eating monarch butterflies, so the more palatable viceroy butterfly wisely mimics its distant cousin's color and pattern. Some katydids conceal themselves as unappetizing leaves, their wings simulating rot spots and chewed areas.

Conspicuous deception is the strategy of still other insects. The pupa of a Burmese moth resembles the head of a bird-eating reptile; certain flies are made in the image of honeybees; the back end of the tortoise beetle displays a frightening red face with golden eyes, and the pupa of the spalgis butterfly mimics a monkey's face; the alligator bug of Brazil duplicates its namesake reptile in miniature— including a deadly looking, if illusory, row of sharp teeth.

Dignity must sometimes be surrendered to disguise. Grayish white, slightly curled, and stained with spots of black, the *Stenoma algidella* moth assures its own safety by appearing indistinguishable from bird droppings. □

Disguised by dead termites and pieces of its victims' nests, an assassin bug lies in wait for unsuspecting prey.

Ambushed

No bushwhacker is quite as attractive as the pink flower mantis of Southeast Asia. Hanging inverted below a twig, this innocent-looking insect displays its pink underside as if it were a beautiful flower. Other insects are lured in and make a feast for the mantis.

Less delicate but no less effective as a trapper is the 200-pound alligator snapping turtle, which lurks in the mud on the river bottom, its mouth agape. Within, a pink flap of flesh undulates at the end of the turtle's tongue, looking like a tasty worm and drawing in unsuspecting fish.

Two snakes engage in a similar trick with their tails: The young copperhead conceals itself under leaves, wiggling the brightly colored tip of its tail to entice frogs; the sand boa, a native of the deserts of North Africa and the Middle East, slips under the sand, exposing only the tip of its squirming tail to lure lizards and rodents within striking distance.

The larva of the ant lion uses no bait, depending instead on its prey to wander into the trap. The creature digs a two-inch-deep pit of fine, sandy soil, then conceals itself at the bottom until an ant ventures onto the slope. Few ants are able to scramble over the ever-crumbling edge of the sandy pit, but the ant lion hastens their demise by scooping away sand below its victims, causing them to slip quickly toward the bottom and into the lion's powerful jaws. □

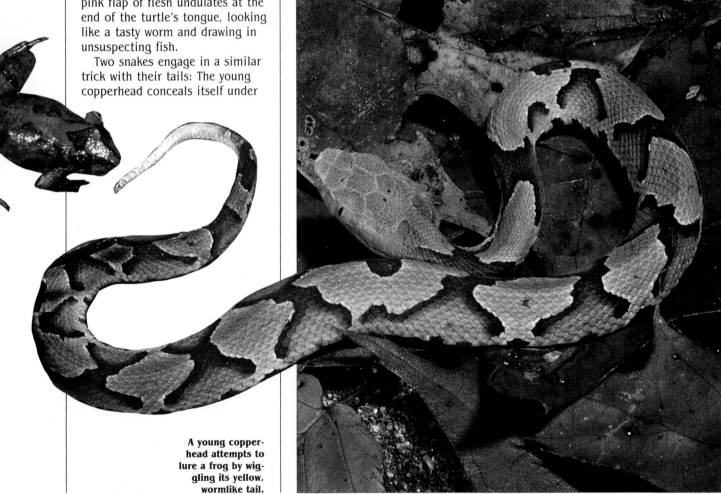

A young copperhead attempts to lure a frog by wiggling its yellow, wormlike tail.

Hiding in Plain Sight

The distinctive black and white stripes of the zebra constitute a paradox of concealment: Although they make the zebra quite visible to the hungry lions and hyenas with which it shares the African plains, the stripes are also the zebra's best defense.

Since predators prefer to attack solitary animals and avoid groups, zebras find safety in numbers, herding together when threatened and galloping away from danger. Their markings are well suited to such a defense, for their repeating stripes blur distinctions between individuals and make it difficult for an attacking carnivore to know how many animals it is facing.

Even when a panicked herd scatters, predators often are confused by the bounding chaos of zebra flight. Narrow stripes decorate the animals' flanks and wider stripes cover their haunches. Randomly exposed during the zebras' typical zigzagging escape, the stripes disrupt predators' perceptions of speed and distance. Only in haze or at dusk—when carnivores prefer to hunt—do the zebra's stripes serve as traditional camouflage. In the soft, gray light, the irregular bands of black and white blend indistinctly with the shapes of grass and bushes. □

Rest in Peace

The sexton beetle is a ubiquitous member of the forest-floor community, plying its nocturnal trade with diligence and anonymity. Perhaps appropriately so, for the sexton, also called the burying beetle, is the undertaker of the animal kingdom, quietly disposing of the bodies of songbirds, shrews, field mice, and other small creatures. Sniffing out the recently deceased, the little morticians—the largest of the species measures only an inch—first inspect the remains, then begin the slow procession that moves the animal to its final resting place. Males and females often work as couples, the male wedging himself beneath the carcass on his back, supporting the body with his six legs and pushing it forward while the female pulls from above and clears obstacles from the path. Their objective, achieved in half-inch increments, is softer ground, where the burial will take place.

There, the beetles excavate earth from beneath the body. Chewing through roots and clipping entangling grass, burying beetles methodically eliminate obstructions in their way. Ultimately, the grave may be eight inches deep; as the body slowly sinks into the hole, the loose soil accumulates on top.

Unlike human morticians, sexton beetles do not stop with proper interment; their goal is food and a nest. Underground, the couple removes the animal's fur or feathers and kneads the carrion into a ball. Dining on putrefied flesh and blowfly maggots, the beetles build their nest in the grave and nourish their hungry yellow grubs in the buried body until they, too, grow into capable body snatchers. □

Beginning at upper left, a sexton beetle readies a shrew corpse for burial. As the beetle digs a hole under the rodent, the shrew slowly disappears into the earth.

Call Me Honey

When the small African bird known as the honey guide begins to chatter at all who pass by—whether mongoose, baboon, or villager—a conspiracy of the appetite is in the air. Those animals in on the secret follow this woodpecker kin as it flits through the trees and calls repeatedly to invite its guests. The honey guide, true to its name, is leading the way to a sweet treat: a nest of wild honeybees.

Although the bird welcomes all comers to the table, its favorite guest is the ratel, or honey badger, a twenty-five-pound bundle of furious energy whose strong claws and muscular jaws can shred a hive in seconds. Protected by a rubber-tough skin that is virtually impenetrable by bee stings, the badger gorges itself on the honey. Its feathered guide patiently awaits a turn at the leftovers—beeswax and grubs—which serve as the main course for the guide's own meals.

The honey guide is unique in its ability to digest the wax, possibly doing so with the aid of intestinal parasites. Its acute sense of smell is unusual among birds; it enables the honey guide to detect bees or the scent of wax at great distances. One sixteenth-century Portuguese missionary living in Africa complained that honey guides swarmed the altar whenever he lit liturgical candles, which are by custom made of beeswax. □

A slip of the tongue is rare among felines, from the most playful household tabby to the wildest king of the jungle. All cat tongues are as rough as sandpaper—so they are excellent tools for grooming fine fur and ideal for scraping the tiniest morsel of meat from the bones of a cat's prey.

Gone Fishing

Not content to simply spin a web and wait for a meal to stumble in, the female bolas spider prefers the sporting life—in this case, fishing for its food. Crouching on a trapeze it spins from a branch, this nocturnal arthropod of the Americas angles for large male moths. It attracts the males to its hunting ground by releasing the sexual scent of a female, which the spider is able to manufacture just for this purpose. Then the spider casts an elastic silk line terminating in a sticky ball, which it swings toward its quarry in the hope of catching the victim in midair. When it gets a strike, the spider reels in its prey or scurries down the line to dine.

Another spider, the angler, often fishes for the real thing. Although it usually eats insects that fall into the water, the angler also lures tiny young fish by rapidly tapping the water with an extended leg, almost the way a fly fisherman dabs the surface to attract trout. As the fish approaches, the spider dives after it, eventually emerging with one that did not get away. □

he water spider of Europe and
sia lives a submarine life, hunting
sects and freshly hatched min-
ows, molting, mating, and even
aying its eggs while submerged.
ts home is a teacup-size tent, anchored to vegetation that grows
below the surface of lakes and
streams. To inflate this silken diving bell and replenish its air, the
spider periodically swims to the
surface, gathering air in bubbles
captured between the hairs of its
abdomen and legs.

Odd Couples

During India's dry season, when
the low-lying pastures in which
they normally forage stand parched
and bare, herds of small spotted
deer called chitals retreat to the
shade at the edge of the forest. In
the branches above them live
troops of long-tailed hanuman langurs, whose finicky eating habits
provide the chitals with food—a
steady rain of leaves plucked from
the trees but rejected and discard-
ed by the monkeys. It is a rain of
plenty: A single troop of twenty
langurs may drop 1.5 tons of
leaves to the ground each year. ◊

**Mutual preservation pairs African buf-
falo and oxpeckers. Parasites on the
buffalo sustain the oxpeckers while the
birds act as lookouts.**

The cooperative relationship between monkey and deer extends to mutual protection. The chital uses its keen sense of smell and the langur its sharp eyesight to watch for the animals' mutual enemies, the tiger and the leopard, and each understands and heeds the other's alarm call.

Similar instances of mutual aid are uncommon in the wild, and when they occur they may sound more like fables from Aesop, not facts of natural life. For example, ostriches often graze with antelope and zebra on the African savanna, feeding on insects turned up by the hooves of their hosts. With their superior eyesight the birds scan the horizon and sound a warning at the approach of danger.

In wild partnerships, hygiene, housekeeping, food, and security are the goals. The African tick bird rides the back of the rhinoceros, the buffalo, the zebra, the warthog, and other large animals. Feasting on bloodsucking ticks embedded in the animals' hides, the birds also watch for and warn of predators. The spur-winged plover, heedless of any danger, flits in and out of the open mouth of the basking crocodile, dining on the leeches and other parasites that otherwise would injure the reptile. □

LOVING AND LIVING

To find a mate, to bring up offspring, to overcome the obstacles of environment: These are the imperatives that dictate patterns of life in the animal kingdom. To these ends, wild creatures become flamboyant exhibitionists and builders; they make enormous sacrifices and show unexpected tenderness. For some, nature dictates devotion to their mates and the young; for others, cold-blooded animosity to their own kind.

Often, animal homes are works of art and feats of engineering prowess. They may house a single creature, a family, or a colony—even serve as an environment for many species.

Surprisingly, though, the most elaborate structures are not homes at all, but colorful creations whose purpose is the most elementary and essential of all in the wild race for survival: to attract a mate to give birth to the next generation.

Father's Work

In the howling blackness of the antarctic winter, all life must struggle to survive. Yet few creatures struggle more than the majestic emperor penguin, which raises its young on the barren winter ice, braving 100-mile-per-hour winds and temperatures that plummet to seventy degrees below zero Fahrenheit.

To succeed at the task, penguins adopt a novel and sophisticated family lifestyle: The male stays home, and the female treks off in search of food, a journey that takes two months and covers more than 100 icy miles.

The extraordinary effort of raising young under such conditions comes about because of a peculiar necessity: Penguin chicks must be hatched in the middle of winter if they are to grow big enough to survive the following winter on their own. And the handsome, golden-bibbed, black-and-white birds, which stand about four feet tall and weigh up to 100 pounds, must establish their rookeries far from the ocean's edge to ensure that their homes will not melt away during the summer thaw. Each April, at the beginning of antarctic winter, tens of thousands of penguins rendezvous at the nesting site. Couples, which mate for life, find each other in the raucous throng by trilling their own unique little song. Together again, they stand breast to breast, throw their heads back, and sing out loud. After mating, the female lays a single egg in June, then leaves on an odyssey that will take her to the ocean and back. Her task is to eat and fortify herself, and to carry back in her crop—an internal pouch located off the esophagus—enough food to feed the chick. Antarctica is now gripped by winter; the pack ice is growing wider by the hour, and the ocean may be 60 miles away. Yet if she should fail to return, the family will perish.

The father remains at home coddling their egg. There is nothing on the ice with which to build a nest; instead, he nestles the egg on his feet and covers it with a warming flap of skin on his belly.

Steam rising from the heat of their packed bodies, emperor penguins *(above)* huddle against the antarctic wind. At left, an emperor chick peers out from between its parent's legs.

The patient male eats nothing during the sixty-four-day incubation period; instead he relies on stores of body fat to keep him alive. He can lose half his body weight during this fast.

To stay warm, he huddles with other egg-sitting male penguins in a tight pack, snuggling bill-to-back in a group of up to 6,000 birds. The huddle is remarkably effective; the temperature in the middle of the pack can be raised by as much as twenty degrees Fahrenheit, reducing the birds' heat loss by 25 to 50 percent.

The birds on the outside of the huddle bear the full force of the wind. So that all may survive, the penguins take turns as windbreaks; the entire huddle swirls slowly as those on the inside and outside trade places. After two months, the almost-naked chick hatches and remains in the safety of its father's brood spot. If the mother has not yet returned, the father feeds his newly born offspring with "penguin's milk"—not milk at all, but a nourishing secretion from the penguin's esophagus.

When at last the mother returns, she is carrying up to 7 pounds of fish and squid in her crop. The chick is transferred to the female's feet for care and feeding, and the emaciated male begins his own trudge toward the ocean to bring back fresh supplies of food for their voracious hatchling. □

Strutting Their Stuff

The bird of paradise, which lives in the forest of New Guinea, sports such spectacular plumage that when sixteenth-century explorers first took some bird skins back to Spain, the Europeans believed the magnificent creature could live only in heaven. Indeed, no other bird's feathers display such variety of color and form, from lacy plumes to metallic, vanelike flags. The glory of their plumage is used by males to impress potential mates, but they don't need females around to strut their stuff. Two plumed males will perch in a tree and dance together—bugling, bowing, and raising their feathers in step—for an audience consisting largely or wholly of other males.

The dancers often resemble blossoming plants more than birds, and their inventiveness is astonishing: One species performs upside down; another clears a stage on the forest floor that can measure fifteen feet across. Still another spends hours plucking leaves from the trees above so a shaft of light will illuminate its magnificence to best effect. □

Hanging inverted below his intended mate, a male bird of paradise displays his plumage.

The Eyes Have It

Early explorers of the wilds of Sumatra and Borneo recorded encountering an extraordinary sight: Now and then, in a jungle clearing, they would find themselves surrounded by an undulating circle of shimmering, three-dimensional spots floating in the air like hundreds of iridescent eyes.

This was the courtship display of the great argus pheasant, which mounts one of the most splendid visual demonstrations in the animal kingdom. In everyday life, the great argus pheasant looks impressive enough, trailing a long, colorful plume of feathers behind its chicken-size body. But during courtship, the already imposing male is transformed into a living work of art.

After clearing a dance area on the jungle floor, the male unfurls his wings overhead, so that his head peers from an enormous circular ruff. In this position, the angle of light changes the hundreds of spots on the bird's wings from two-dimensional tan circles into startling three-dimensional eyespots, which scintillate like the eyes of a host of dancers. Scientists speculate that the spots excite females because they resemble kernels of grain, which males of some other pheasant species offer to their intended. □

Love Nests

The bowerbird of Australia and New Guinea is the master builder of birds. When the time comes to woo a mate, the male—who may be no bigger than a robin—builds an elaborate love nest that occasionally stands seven feet high. It is a magnificent gesture intended to attract not one, but a procession of mates.

Romance, bowerbird style, starts when the male clears an area three to five feet wide, then surrounds it with a construction of thousands of twigs, an effort that can take weeks. One species of bowerbird, only nine inches long, constructs a ''maypole'' tower up to nine feet tall by twisting twigs around a young tree in the middle of the dance floor. Another species pre-

fers a round hut just a foot high. Some birds engineer fenced avenues, carefully laid out north-to-south in order to gather the brightest of the sun's rays. One constant prevails, however: The plainer the bird's own plumage, the more spectacular the bower's construction.

When building is completed, the bird decorates his love bower with visual enticements, natural and artificial. Leaves and iridescent insect casings join bottle caps, bits of paper and glass, hair curlers, and other human castoffs. Australians have found missing car keys by searching nearby bowers. Bowerbirds can accumulate substantial treasure: One display featured more than 500 bleached kangaroo bones and 300 snail shells.

Yet the birds do not necessarily

equate quantity with quality. They are finicky decorators with decided preferences for certain colors and materials. Some bowerbirds paint their constructions, mixing berries and other vegetation with charcoal and saliva and then spreading the mixture with their bills.

His creation finished, the bowerbird stands outside and calls loudly. Ironically, though, a female who is attracted by the builder may not care how elaborate the decorating scheme. If she is ready to mate, she will enter a humble bower as readily as she will a grand one. The final irony: Her eggs will be hatched in a very ordinary tree nest nearby. The polygamous bower-building male, meanwhile, turns his attention to the task of refurbishing his structure in order to attract other mates. □

A male bowerbird indulges a passion for blue by decorating with other birds' feathers.

Tapered Eggs

Seabirds such as the murres and the guillemots nest on narrow cliff ledges high above the ocean. The precarious nature of their homes has led to a peculiar, but useful, adaptation: Their eggs are more sharply tapered than those of their inland brethren, so that they roll in tight circles, lessening the chance that they will fall off a ledge. By contrast a chicken's egg rolls in an almost straight line. □

Walking on Water

The western grebe literally walks on water for its mate. In their stylized courtship ritual, pairs of these North American water birds lurch out of the water and skitter frantically across the surface, their feet paddling furiously amid a cloud of spray. This remarkable aquatic performance is known as rushing.

The rushing of romantically inclined grebes is accompanied by a graceful synchronized water ballet, in which the birds dive and surface in unison, concluding the dance by rising from the water, each holding a piece of weed in its beak. Bills pointing to the sky, they join the weeds and hold them aloft, as if symbolizing their union. Scientists have yet to unravel the mystery of the grebes' dance: The eminent American naturalist Aldo Leopold confessed himself "helpless to translate" its "secret message." □

Tying the Knots

The fuzzy balls on the branches of trees in Africa and Asia are not holiday ornaments, but one kind of nest built by a species of the industrious weaverbird. The nests, built only by males, serve as a stage prop for courtship and as a home after mating.

The masked weaverbird first uses fresh grass or strips of palm fronds to weave a perchlike ring on a tree branch. Sitting in the ring, he then weaves and knots grass to make a sturdy shell around himself. Some birds even use slipknots, overhand knots, and half hitches. The birds start practicing building skills in their youth, for their craft improves with repetition.

The practice is important: If a female is satisfied with a nest, she mates with its builder, but failure dooms the male to demolishing his handiwork and starting anew. Some weaverbirds are polygamous, and they can be very busy indeed; the male may make three or four nests in succession, one for each of his mates.

The species known as sociable weaverbirds, which congregates in groups of up to 250, collaborates on apartment-style nests up to fifteen feet across *(below)*. The nests can become so heavy that they break the branches supporting them, forcing a move to a new neighborhood. □

Food for Love

Many animals seek to tempt the object of their affections by offering tasty tidbits.

For example, the male black-tipped hangingfly of North America attracts a mate by dangling a plump housefly before a female. But if the love offering is too small, the male is rejected; the gift must contain adequate protein for egg production. If the gift is satisfactory, the two flies mate while the female dines.

The dance fly of New Zealand bears gifts of food to avoid becoming food: The female of the species will eat her mate if given half a chance. To escape such a fate, the male brings along an insect when approaching a group of females; when one of them grabs the food, the male grabs her, whisks her off to a nearby bush, and accomplishes his mission while she devours the cuisine.

Some dance flies are deceptive. They gift-wrap their presents in strands of silk, often concealing the fact that the gift is nothing more than a bit of vegetation rather than a tasty morsel. By the time the female discovers the ruse, the male has departed. □

Nesting Instincts

Birds are not the only creatures with a nesting instinct: Harvest mice, tiny denizens of the fields, weave aerial nests supported

by a few slender stalks of wheat or tall grass *(right)*.

Using its teeth and delicate paws, the mouse shreds grass into thin strips, which it weaves into a ball; the males make loose nests for sleeping, and the females make sturdier nests, lined with thistledown and cattail fluff, for raising their young.

Within three weeks of birth, the baby mice become so rambunctious they can tunnel through their handsome home, tearing it to shreds. From that point on they are on their own. □

Married for Good

There is no wild animal that stays paired with a mate longer than the jackdaw, a bird that gives human beings a run for the golden anniversary award.

Although jackdaws are highly social, they are monogamous. Once these affectionate birds pair off at about the age of one year, they remain faithful for the rest of their lives, which last as long as sixty-five years. □

Love Gone Wrong

Unlike mammals, birds do not instinctively recognize others of their own species. Instead, baby birds learn their identity by a process known as imprinting.

Usually, imprinting works through baby birds' close association with their parents. However, imprinting can go wrong. Konrad Lorenz, the famed Austrian naturalist, found that after he crept around his garden quacking to a brood of orphan ducklings, they thought he was "mother" and followed him everywhere.

The same perverted process ruined the adult love life of an aging peacock that once frequented the grounds of the Palais de Nations in Geneva, headquarters of the World Health Organization. Somehow, the bird became enamored of the organization's noisy, smelly, hand-cranked mimeograph machine; as WHO employees churned copies from the machine, the peacock perched on the window sill and spread his tailfeathers in full courtship position. He ignored the many charming peahens that were offered as mates and died a confirmed bachelor. □

Tied Up

The female European crab spider dons a "bridal veil" during courtship. But the purpose is not to beautify the bride; the veil protects the groom from his mate.

Female spiders often are larger and stronger than the males and have the alarming habit of eating their beaux after mating. The male crab spider avoids this grisly fate by actually tying up the female with a veil-like web, which he spins during courtship. By the time the female has wriggled free, the male has finished his procreative duties and is long gone. □

A male crab spider *(top)* massages his mate-to-be while tying her down with his web.

Organic Incubators

Mallee fowls hatch their eggs not by sitting on them, but by building a heated incubator. This engineering marvel regulates the temperature of the nest to within one degree of ideal, thus protecting the fowl's eggs from the harsh climate of the central Australian plains where the mallee makes its home.

Such an engineering enterprise comes at a price: Maintaining the system keeps the bird working eleven months of the year. Construction starts in April or May, when a mating couple digs a pit almost a yard deep. Then they fill it with leaves and twigs and cover all with a mound of sand. The result is a compost heap whose temperature gradually rises as the vegetation decomposes. After about four months, the heap warms to an ideal 93.2 degrees Fahrenheit, and egg laying begins.

For the following six to seven months, through cycles of egg laying and incubation, the industrious male maintains the nest's temperature, performing temperature checks by probing the mound with open beak and then digging ventilation shafts or adding insulating sand as temperature changes dictate. Even when scientists have surreptitiously warmed a mallee fowl mound with an electric heater, the birds were still able to keep the temperature constant. □

Family Fortress

The hornbill's gigantic beak is prized by artisans of Southeast Asia, who carve complex designs similar to whalers' scrimshaw in the oddly shaped, ivory-like material. But to the hornbill, the beak is far more than merely decorative; it plays a vital role in the bird's strange nesting habits.

The female hornbill spends her breeding time literally walled up in the hollow of a tree, with the entrance almost completely closed by a mud barrier built by the female and her mate. Safe from predators, she incubates her eggs while the male provides as many as thirty meals a day, pushing food through a narrow slit in the wall. For much of her "imprisonment," the female's escape from her fortress would be fruitless, for she could not fly; while on the nest, she molts her feathers completely, behavior rare among birds.

When the eggs hatch, the male's feeding trips become twice as frequent—as many as seventy every day. When the offspring's appetites become so hearty that the father's efforts cannot keep pace, the female breaks out of the nursery, battering away the mud wall with blows from her massive beak, and joins in the food gathering.

However, no sooner does she breach the wall than the young birds plaster it back up again; they remain happily contained by the mud barrier until, at about six weeks of age, they feel big enough to stage a second breakout and set off on their own. □

The hole in the mud
wall that protects
female and chicks
allows an ornately
beaked male horn-
bill to pass food
to his mate.

JOSHUA THE MOOSE nuzzles Jessica, a Hereford cow to which he became attracted when he wandered onto a farm in Shrewsbury, Vermont, in 1986. The unfulfilled romance, which attracted worldwide attention, came to an end after seventy-six days, when the moose's antlers—and his libido—dropped off.

Mateless

The female of many species of whiptail lizard begets only females—each daughter a clone of the mother—without benefit of a mate. Indeed, males do not even exist in one-quarter of the lizard's forty species. Scientists think this unique reproductive behavior may help the lizards survive in their harsh desert habitat by avoiding a time-consuming search for a mate. □

Shore Leave

Elephant seals live two-thirds of their entire lives in the sea. And when these 5,500-pound creatures clamber onto dry land, it is for a wild, four-month beach party of brawling, birthing, and mating.

In early December, the males come ashore on islands off southern California and Mexico and start fighting, burping, bellowing, and gouging each other with their powerful canine teeth. They make a frightening commotion, and their teeth inflict bloody wounds, but their encounters are not life-threatening. However, the outcome of the battles is crucial: The losers have no chance to mate, and up to 90 percent of the males can be excluded. In fact, some elephant-seal males never mate during their twelve- to twenty-year lifetimes.

After the males have established the social order, the females arrive, and within a week those who mated the previous year give birth. The pups, which can be five feet long, gain as much as twenty pounds a day during their first month of life, while the nursing mothers lose several hundred pounds during the same period.

Childbearing completed, the females congregate in harems of about forty, each under the watchful, jealous eye of a dominant male, who takes on a frenetic schedule of mating and guarding his consorts, with rarely a moment to sleep. Throughout the experience, neither males nor females eat, for no food is available. Finally, after the babies are weaned and the adults have shed their old fur, the entire troop returns to the calmer confines of the ocean. But some will remain behind: A few bulls die, exhausted by the efforts of the past four months. □

Elephant seal bulls wage a bloody battle for mating rights on California's Farallon Islands.

Swift Work

Some swiftlets—small, quick-flying birds that dart over the forests of tropical Asia—have concocted a novel solution to the problem of securing their nests to overhanging cliffs or cave walls: They make the nests with spit.

When breeding time approaches, the swiftlets' salivary glands swell to several times their normal size. Both male and female work at nest building, beginning by tracing the semicircular outline of their nest on the rock with their tongues. After more than a score of passes, the thick, quick-drying saliva has built a small translucent cup, in which the birds' eggs are laid. The practical swiftlets also use saliva to glue together little balls of insects, forming an ideal snack for their young. Unappetizing as it may sound, swiftlet nests are the chief ingredient of bird's-nest soup, long prized as a delicacy by oriental gourmands. □

Animal Midwives

Animal births may take place far from prying human eyes, but they are hardly solitary affairs. Instead, many creatures of the wild, from great elephants to tiny mice, help each other through the difficult birthing process.

When an elephant is about to give birth, she retreats into the bush with one or two other females from her group. They guard the expectant mother while she is in labor and help keep her on her feet. Sometimes the midwife elephants use their trunks to assist in the delivery; afterward, they help the newborn elephant to stand, nudging it with their feet and trunks. Once the baby is up and about, young female elephants act as affectionate and fascinated babysitters.

Some of the best animal midwives are found among rats and mice. Egyptian spiny mice, for example, act as midwives in two-thirds of all deliveries, assisting in delivery, biting the umbilical cord, and licking the babies clean while the mother continues delivering the rest of the litter.

Animal midwives are almost always female. One exception, however, is the male southern bush rat of Australia. This unusual helpmate has been observed aiding his mate by licking the babies and keeping them warm in the nest. □

Murderous Deceits

For birds as well as humans, raising children is hard work. But some cuckoos—most notoriously the European species—avoid the difficulties of child rearing by recruiting foster parents, using techniques of murderous deceit that would do the most diabolical mystery writer proud.

At egg-laying time, the female cuckoo—who never bothers to build a nest of her own—swoops down into an unguarded nest of another species and lays her egg. Some compound the crime by flying away with one of the other bird's eggs and promptly destroying it. If the cuckoo lays more than one egg, each will be deposited in a different nest. Sometimes the unwilling host rejects the interloper's egg, but most frequently the cuckoo egg's shape and color are acceptable to the host, and the nesting bird will nurture it. Cuckoos cleverly lay eggs that help meet this color requirement—European cuckoos have both spotted and plain blue eggs—and seek out a host to match.

When the cuckoo egg hatches, murder follows deceit. The cuckoo chick emerges one to four days earlier than its foster siblings, and within ten hours of hatching, the naked, blind baby cuckoo begins to scoot around the nest; when it encounters another egg, instinct prompts the bird to laboriously pry and push it over the edge to certain death. This fierce expulsion instinct persists for several days and usually succeeds in ridding the nest of all but the rapaciously clever cuckoo.

The cuckoo's parasitic ploy is accompanied by a huge hunger; in order to sustain its rapid growth rate, a cuckoo chick easily consumes all the food that was intended for the brood it killed. More often than not, the hapless foster parents raise the interloper as their own—unaware that it has killed their own offspring—even though the enormous cuckoo chick may tower over them, completely filling the nest. □

A cuckoo chick dominates the nest of a dunnock, whose own eggs may have been destroyed by the chick and its mother. Despite the cuckoo's great size, the dunnocks continue to raise it.

Being born is never easy on babies, but giraffes have an especially tough trip: Mother giraffes give birth standing up, and their infants drop six feet to the hard ground. Amazingly, the 160-pound newborn giraffe is not hurt by its rude arrival. With its mother's help, the sturdy child is standing up within five minutes and enjoying its first meal within twenty.

A Little Bit Pregnant

The female armadillo—almost alone among mammals—is capable of being just a little bit pregnant. These odd armored mammals have the ability to delay the development of a fertilized embryo for up to three years after mating, waiting until food and habitat are right for producing offspring. Once the pregnancy is begun in earnest—when the embryo is implanted in the mother's womb—it is only three months before four identical offspring are born.

The armadillos' odd reproductive strategy, which ensures that the young will not be born in times of drought or food shortage, has enhanced the species' survival and spread. Although the creatures came to the United States only 100 years ago, they now range over most of the southern half of it. □

A mother armadillo nurses her week-old identical quadruplets.

Group Motherhood

Bears are normally not sociable creatures. But mother bears sometimes cooperate with each other in raising their young. During this long and intensive process—it takes two years for the eighteen-ounce newborn to grow and learn self-sufficiency—mothers babysit and nurse each other's children, trade information about food sources and good spots to den, and even exchange or adopt cubs.

Adoption is particularly important. A bear's survival skills are learned, not inherited, so orphaned bears often do not have the skills they need to feed themselves and so starve to death. □

Swallowed Up

The female gastric-brooding frog of the Australian rain forest swallows her own eggs—not for nourishment but to safeguard them from predators.

Never eating while the eggs are in her stomach, she holds them there while they hatch into tadpoles. She then retains the growing tadpoles even when they become so large that they collapse their mother's lungs, forcing her to absorb oxygen through her skin.

Only after several weeks, when her offspring have turned into fully developed little frogs, does the mother regurgitate them into the perilous world. □

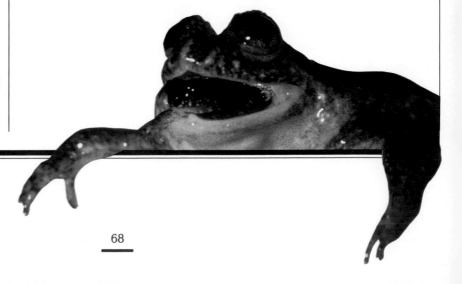

Cain and Abel

Great egrets don't know the meaning of the phrase "brotherly love." These handsome white birds regularly kill their siblings while they are still in the nest, all the better to survive themselves.

This brutal self-interest is common among egrets and other wading birds, as well as some birds of prey. All hatch their eggs at one-to three-day intervals. When the younger siblings emerge from the shell, the larger and stronger earlier-born chicks immediately begin pecking at them and stealing their food. The unrelenting attacks have deadly consequences; one study concluded that about half of the younger nestlings die during their first month.

Scientists speculate that such behavior is actually a survival device for the species that engage in it. If food is plentiful, all the offspring could survive, but if food is scarce, the parents are assured of having at least one survivor to carry on the bloodline. □

Near its seemingly unconcerned parent, the older egret chick on the left strikes a heavy blow to its sibling.

Extra Large

Bearing children is never an easy job, but it is especially difficult for a mother kiwi. The eggs laid by these flightless New Zealand birds are five inches long and weigh in at a full pound—one-quarter the weight of the mother bird. Ornithologists say that eggs this size would ordinarily be laid by a bird six times as large as the kiwi.

The eggs are so huge that the hen-size kiwi must waddle, legs splayed wide, for several days before squeezing out an egg. The hapless kiwi often lays two and sometimes three eggs in a clutch, spacing them about one month apart. The male kiwi takes on the arduous task of incubating the eggs by draping his entire body across them.

To be sure, the extra-jumbo egg serves a worthwhile purpose. It provides so much nourishment that the young birds hatch fully feathered and require little care from their parents. □

An x-ray of a female kiwi reveals the outsize dimensions of her egg one day before laying.

Pouch Potatoes

The female kangaroo has the world's finest baby carrier: her pouch. But far from being just a handy accessory, the kangaroo's cozy container is a necessity.

Most infant mammals become highly developed in the mother's womb before birth, but baby kangaroos come into the world after only five weeks of gestation, still only the size of a bumblebee. The deaf, blind, and hairless baby—called a joey—must make a perilous journey up its mother's abdomen into the special pouch that is unique to marsupials. There, it clamps onto one of four teats, which swells and stiffens to hold the infant firmly in place for several weeks. Because the little joey is too weak to suck on its own, the mother's teat pumps milk into its mouth.

The young kangaroo stays nestled in the pouch for three to nine months as its mother bounds through the Australian bush. Once it emerges from the pouch, the baby will follow its mother for several months, scrambling head-first back into the pouch's safety at the first sign of danger.

The female mates again shortly after giving birth and, as a result, can be dealing with three offspring at a time: one in her pouch, an older sibling behind on foot, and a third in the womb.

To satisfy the different demands of each youngster, the mother kangaroo produces two kinds of milk: high-fat milk for the active older

joey, which pokes its nose in the pouch to nurse, and a protein-rich version for the joey in the pouch. Development of the embryo in the womb is delayed until its older sibling moves out of the pouch. If the infant dies, the embryo will begin developing at once, ensuring faster reproduction. □

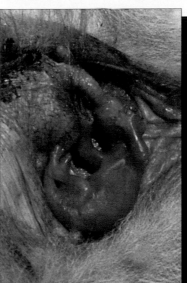

Head and feet emerge from its mother's pouch as a large kangaroo joey *(above)* scrambles to get comfortable. At left, a tiny two-week-old kangaroo clings to its mother's teat within the pouch.

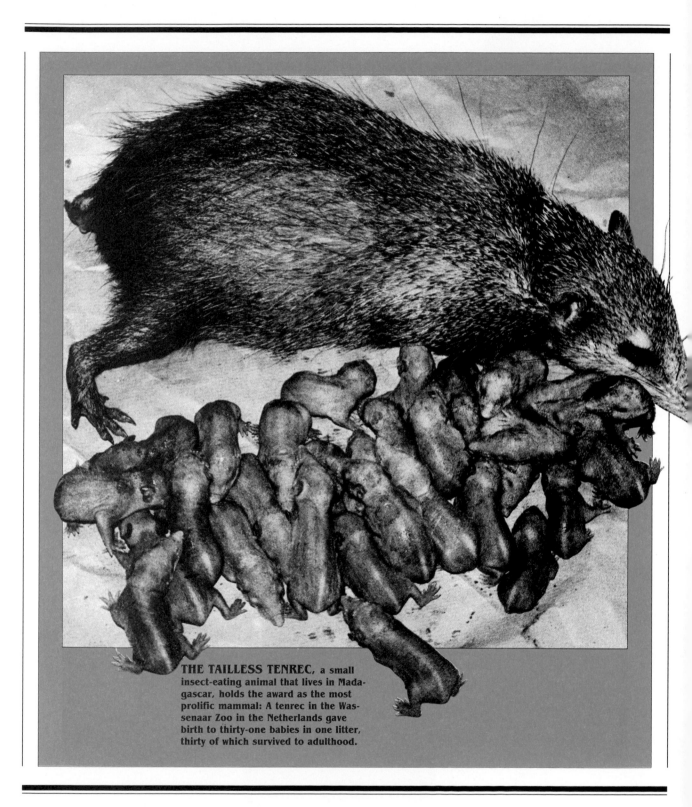

THE TAILLESS TENREC, a small insect-eating animal that lives in Madagascar, holds the award as the most prolific mammal: A tenrec in the Wassenaar Zoo in the Netherlands gave birth to thirty-one babies in one litter, thirty of which survived to adulthood.

Queen Bee

Honeybees have the most elaborate family life of any insect, although virtually every action of a hive's 80,000 members revolves around one bee—the colony's queen, on whom they depend for a continuing supply of workers.

Queen bees start out life genetically no different from any other bee, as larvae growing in a cell in a hive. But certain of the bee larvae live in larger cells and are fed more of a special food called royal jelly, which is produced by nurse bees. These pampered infants are truly anointed, and one of their number will grow up to be a queen; the others will die, for a hive can have only one queen. When the first queen emerges, she quickly kills her sisters in the other queen cells.

Then the new monarch sets off on the one mating flight of her life, enthusiastically pursued by a group of males, or drones. The first drones to reach her mate with her in midair. The privilege comes at great price; each of the males dies. The drones' sperm lives on for years, however, enabling the queen to produce hundreds of thousands of offspring through ◊

Sharp teeth and a keen nose equip the naked mole rat for a life underground.

Queen Rat

The naked mole rat—a wrinkled, pink creature with large teeth that lives beneath the dry plains of East Africa—looks and behaves like no other mammal. Its large colonies are like those of some termites, headed by a queen that is attended by one or two breeding males.

These bewhiskered animals spend their lives underground in colonies of ninety or so. The queen and her courtiers spend their time reproducing. The few males who are so favored pay a price; they shrink to about half their original size after mating. The mole rat queen, however, like queens of insect colonies, may grow to twice the size of her subjects.

The workers, three- to six-inch-long males and females, scuttle around maintaining the colony's elaborate system of tunnels, which can cover an area the size of six football fields. They also gather food from plant roots and play nanny to the queen's offspring, which can number nearly ninety each year. Scientists still have a great deal to learn about the lives of these repulsively ugly creatures. In the meantime, an increasing number of zoos are adding naked mole rats to their collections, simply because they are so grotesque and unusual. □

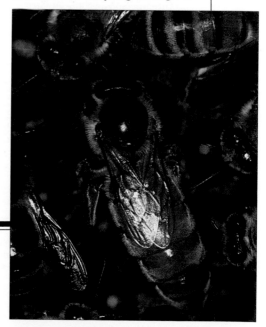

her three to four years of life.

Most of these offspring will become the hive's workers, which are all sterile females. They tend to the many tasks at hand, with each worker taking a specific job according to her age: The youngest clean the comb's cells and incubate the brood; the oldest forage for food and guard the hive.

The queen stays in contact with her busy subjects by distributing a pheromone called queen substance, which is spread through the hive as workers touch the queen and each other. As long as there is queen substance in the hive, the workers know all is well.

A gradual reduction in the amount of queen substance in-forms the colony that the hive is overcrowded and it is time to swarm: A new queen will be made, and some members of the colony will follow her to a new home. But if the level of the pheromone drops suddenly, they know that the queen is dead, and they must quickly begin building cells to raise new queens. □

Paper Castles

Wasps, much feared for their sting, are also admired as architects, building handsome and sturdy homes out of nothing more than paper that they make themselves.

Wasp nests begin with a circular foundation built, like a beehive, of hexagonal cells that are stuck out of the way on a tree or under the eave of a house. Making hundreds of thousands of trips before the colony's home is finished, the female wasps manufacture paper by scraping tiny bits of wood from trees, buildings, or fence posts, and chewing them into a fine pulp mixed with saliva. The wasps smooth the pulp into a sheet with their jaws.

Although the wasps' paper castles are thin and lightweight, they are extraordinarily strong and are capable of housing 200 of the creatures in comfort. Air spaces between multiple layers of paper serve as insulation. In the breeding combs at the heart of the hive, the temperature is kept at a constant eighty-six degrees Fahrenheit by workers whose only function is heating and cooling the nest. They shiver to produce heat, and fetch drops of water to cool the hive by evaporation.

The fine house—the product of so much summer labor—is used for just one season. Only a few young queens survive the winter by hibernating in a sheltered place. In the spring, they begin new colonies in new nests. □

**A work in progress, this paper nest may
grow to support two hundred wasps in
the height of the season.**

Termite Condominiums

Towering as high as twenty feet and housing up to ten million residents, the homes of termites are urban environments within a single structure, condominiums complete with air conditioning. No insects build more elaborate dwellings; if termites were as tall as human beings, their mounds would soar to four times the height of New York City's Empire State Building.

These city-size termitaries are built of feces, particles of wood, or soil and saliva dried as hard as concrete. They incorporate an extraordinarily complex system of channels and air ducts designed to keep the inside cool, moist, and well provided with oxygen.

Even the compass orientation of the mounds is considered by the builders. In Australia, for example, some termites build long, thin structures whose narrow ends face exactly north and south, so that the scorching midday sun, standing high in the northern sky, falls on the smallest area.

Termite colonies are headed by one or two royal couples, which are parents to the entire horde and may live as long as ten years. The ranks of termites are specialized, although, unlike bees and ants, both male and female termites engage in work. Workers tend the king and queen and their eggs, build and repair the mound, and search for food at night. Soldier termites defend the colony.

Inside the termitary, chambers provide living and working quarters, a royal suite for the king and queen, and extensive ductwork for the air-conditioning system.

Although individual termites live only about a year, their splendid constructions are inhabited for as long as twenty years, by generation after generation of industrious creatures. Even after a mound is abandoned by termites, the durable structure may stand for as long as a century. □

This ten-foot-high, concrete-hard home was built by *Macrotermes bellicosus,* a species of termite native to the Ivory Coast.

A Beastly King

Although lions have a reputation as the most courageous of beasts—who would not want to be called lionhearted?—male lions have trouble living up to this particular image.

Males are reluctant hunters, preferring to wait for hyenas or female lions to bring down prey before they amble over, shoulder the others out of the way, and gorge themselves.

Nor are male lions much for family life. Although the king of beasts is the only cat that lives in large groups, called prides, the male bosses of the pride play no role in family life, although they do protect their cubs from harm.

Young males leave the pride at about two years of age and gather in small groups sometimes called gangs. Females usually stay behind, remaining in the pride of their birth for life, raising their cubs communally with their sisters, aunts, and cousins.

By the age of five or six, male gang members are ready to control a pride of their own. Through intimidation and occasional attacks, the gang drives off the old males.

The usurpers then evict the juvenile males and kill the small cubs. There is a biological impetus to the savagery; after losing their cubs, the females become fertile and mate with the new males.

Once in control of the pride, the males revert to the lazy life, lounging and letting the females handle the hunting. Nevertheless, the males must defend their pride against interlopers, a task that exacts a certain toll: Although the life expectancy of a female lion in Tanzania's sprawling Serengeti National Park is seventeen years, males rarely survive twelve. □

Three male lions, each born of a different pride, form the nucleus of a small group sometimes called a gang, whose members will be lifelong companions.

Just Pals

Baboons are among the most promiscuous of primates. Despite this natural disposition, they are capable of platonic friendship between males and females. In fact, female baboons seek to be buddies with certain males.

Females benefit from such alliances in a very practical way; their male pals help care for their infants and protect both mother and child against attackers. But while the relationships may begin as simple friendship, nine out of ten times the once nonsexual pals eventually pair off and mate within a two-year period. □

Busy Beavers

Beavers are among the animal kingdom's most accomplished engineers. The industrious creatures fell trees, build dams and lodges, and create their own private lakes, redesigning nature to give themselves a better habitat.

The beaver is ideally equipped for such tasks. Its thick, oily coat protects it from cold water. Its flat tail stores nourishing fat against lean times and also serves as a rudder. And the beaver's four buck teeth are extraordinarily hard and strong, are self-sharpening, and never stop growing. They can gnaw through a sixteen-inch-diameter tree in just one night.

Like lumberjacks, however, beavers sometimes miscalculate and are crushed beneath the falling timber. But they usually notch the tree so it falls right into the water, where it is easily hauled away. Some beavers dig canals hundreds of yards into the forest so they can float out trees.

Trees, naturally enough, are the chief building material for beaver dams and lodges. Beavers live in family groups of five to eight animals headed by a couple that mates for life. The center of their lives is the lodge. The simplest lodges are chambers dug into a riverbank, connected to the outside by underwater tunnels. If the bank is not high enough for tunneling, the beavers build a castle of branches above ground, cover it with mud, and enter from below.

The ultimate beaver abode is the large midstream lodge, a mound of twigs and mud that can be fifteen feet across with walls three feet thick. Inside is a cozy above-water chamber where the beavers nap together during the day and raise their young.

The dams that beavers build are intended to control the water level and keep the lodge entrances securely hidden underwater. They are enormously sturdy structures, built on a foundation of heavy sticks rammed into the streambed, then piled with branches plastered with mud. The entire affair is anchored, where possible, to nearby trees and boulders. It is not unusual for beaver dams to block the entire flow of a stream and flood the surrounding lowlands. The flooding creates a habitat for many other animals, among them waterfowl, fish, insects, and even moose, which browse on aquatic plants. Beavers use trees for food as well as for engineering, particularly in the winter, when the animals survive by munching on branches they have anchored in heaps underwater. These food caches add to the bulk of their dams and lodges, and can measure as high as five feet and as long as sixty feet. □

A beaver and its kit find warmth and safety inside their lodge, built out of mud and sticks.

TRAVELERS

Animal life is often marked by periodic travel and migration; the change of seasons triggers a search for renewal, for food, or for a mate. So urgent are the impulses that some migrants push themselves to the brink of physical destruction year after year. They cross thousands of miles of ocean, ice, mountain, and desert. Yet the mechanisms that guide the way—so uncanny that they persist from generation to generation, without being taught—remain dimly understood.

Butterfly Trees

A steady stream of orange-and-black wings skitters south each autumn, flowing across eastern Canada and the United States into a remote, densely forested mountain cove in Mexico's Michoacán State, west of Mexico City. The wings flutter into a stand of fir trees within the cove, and when they are still, the branches of the trees sag under the weight of 16 million monarch butterflies. A four-acre patch of forest has been transformed into a sanctuary of butterfly trees, an illuminated cathedral whose orange columns spiral up to the sky *(right)*.

The migration of the monarch is one of the most unusual in the animal kingdom, for the butterflies arriving in the secluded shelter of their Mexican grove in the autumn are the great-great-grandchildren of those that left six months earlier. Yet somehow they find their way and alight on the same trees festooned by their short-lived ancestors. The Mexican butterfly grove, while among the most spectacular, is not unique. Other such groves exist in California, Florida, and Texas, providing winter shelter for the monarchs of the western United States and Canada. Some groves are major tourist attractions, and the northern California town of Pacific Grove, which bills itself as "Butterfly Town USA," imposes a fine and prison term on anyone convicted of killing monarch butterflies.

In March, the monarchs leave the groves on a northward journey whose course is determined by the location of milkweed plants, the only foliage on which the butterfly larvae can survive. Just before they

die, the females of three, four, or more generations—spanning spring and summer—lay tiny green eggs on the plants. The eggs hatch into colorful, tiger-striped caterpillars and soon become the glorious orange-and-black monarchs that carry on the species' migrations.

Researchers speculate that the monarch navigates using its keen sense of smell, which leads the butterfly to the milkweed plant and to the trees of home groves, where the odor from male scent glands may linger, like a guiding beacon left by monarch ancestors. □

Return to Ascension

Ascension Island is a barely distinguishable dot on most maps, an eight-mile-long speck that barely breaks the vast gray-green waters of the South Atlantic midway between South America and Africa. Yet every year, hundreds of green sea turtles navigate to Ascension across 1,400 miles of open sea from Brazil, to mate and lay eggs on the same beaches from which they themselves hatched.

The homecoming ritual begins in late December, when the turtles leave their feeding grounds off the coast of Brazil. Less than a month later, they arrive at Ascension *(below)*, where they begin two months of mating and egg laying, during which the 400-pound creatures scarcely eat. Then, their procreative duties completed, the turtles head back to Brazil.

In May, when the parents have completed their homeward journey, the hatchlings emerge and move instinctively toward the water and a long, untutored journey to the coast of Brazil. They will not see their native beaches again until they return to breed as adults fifteen to thirty years later.

The mechanism of the green sea turtle's navigation remains unknown to scientists. At one time, it was thought that the turtles had learned the route literally inch by inch over a period of 40 million years, as Ascension Island and Brazil became separated by the drift of the earth's crustal plates. It was later discovered, however, that the creatures have been using Ascension for only 10,000 years or so.

Some scientists have theorized that the green turtles follow the sun and stars to find their way. Others maintain that the young turtles are carried from Ascension to Brazil by ocean currents and that in their return journey against the current to Ascension they are guided by a scent of the island that is borne by the water. □

SEEKING SAFETY in the sea, newly hatched Kemp's ridley turtles scurry into the foam of the Gulf of Mexico at Rancho Nuevo, Mexico. Strong currents may sweep some of the turtles as far north as New England, yet each April a few hundred return to lay their eggs near Rancho Nuevo—the only place in the entire world that the Kemp's ridleys breed.

The Ancestral Highway

When spring breathes green life into the stark, white arctic world, the boom of thawing ice and the rush of flowing water make welcome music for more than 2 million caribou that have spent the winter sheltered in the scrubby forests of Canada and Alaska.

Small bands of caribou have survived the winter by pawing through the snow in search of lichen on the forest floor. Now, spring promises fresh, rich pastures in the treeless tundra to the north. The caribou, their bodies worn threadbare by the winter hardship, draw purposefully together, preparing for the northbound migration—the first leg of an annual round trip that may encompass 3,000 miles.

The caribou begin by gathering in small groups, later consolidating into larger and larger bands, threading northward along wooded river valleys or along windswept ridges. When the animals emerge from the northern edge of the forest, they become a woolly, brown current, a flood tide of fur and antlers pulsing across the undulating arctic tundra.

Pregnant females set the pace for the herd. Like the timekeeping ticking of a thousand metronomes, the clicking of the caribou's ankle bones counts cadence for the marchers. Eating steadily as it goes, the mass moves with a single mind, following a route trod by countless generations.

There are ten large caribou herds in North America, each undertaking a similar, separate, journey. From first to last animal, each current of migrating caribou—numbering as many as 200,000 to 400,000—may extend 185 miles.

Although observers are impressed with the grandness of the spectacle, for the caribou it is a difficult, perilous journey. Wolves linger behind the herds, attacking stragglers. Indian and Eskimo hunters await, counting on the herd's arrival to feed their families. Ice-filled rivers, swollen by storms and melting snow, sweep other caribou to their death.

In May or June, each of the herds pauses for the ritual of birth at calving grounds scattered across northern Canada and Alaska, sharing the featureless landscape with millions of screaming migratory birds. The journey resumes several weeks later, after the young grow enough to feed on plants in addition to their mother's milk.

When the trek concludes, the caribou settle down to feed on cotton grass, flowering plants, and tender shoots of scrub willow and birch. Still, there is no rest for the wanderers, for insects—mainly warble flies and mosquitoes—descend in thick clouds to feed on the animals' blood. Some estimate that a single animal loses as much as one quart of blood per week during the peak of summer.

Autumn arrives in mid-August. Frosts and snow return to the tundra, and the caribou herds turn south for the winter shelter of the forest. Although fifty pounds heavier, the caribou push south more swiftly; strengthened by the rich summer diet and unencumbered by pregnant females, they travel almost 10 miles a day.

The caribou mate along the fall migration route. When the migrants finally reach the shelter of the woods, the herds disperse and spend the winter in relative solitude, waiting to set out once again for the north. □

Caribou bulls forge the Kobuk River in arctic Alaska during their annual migration.

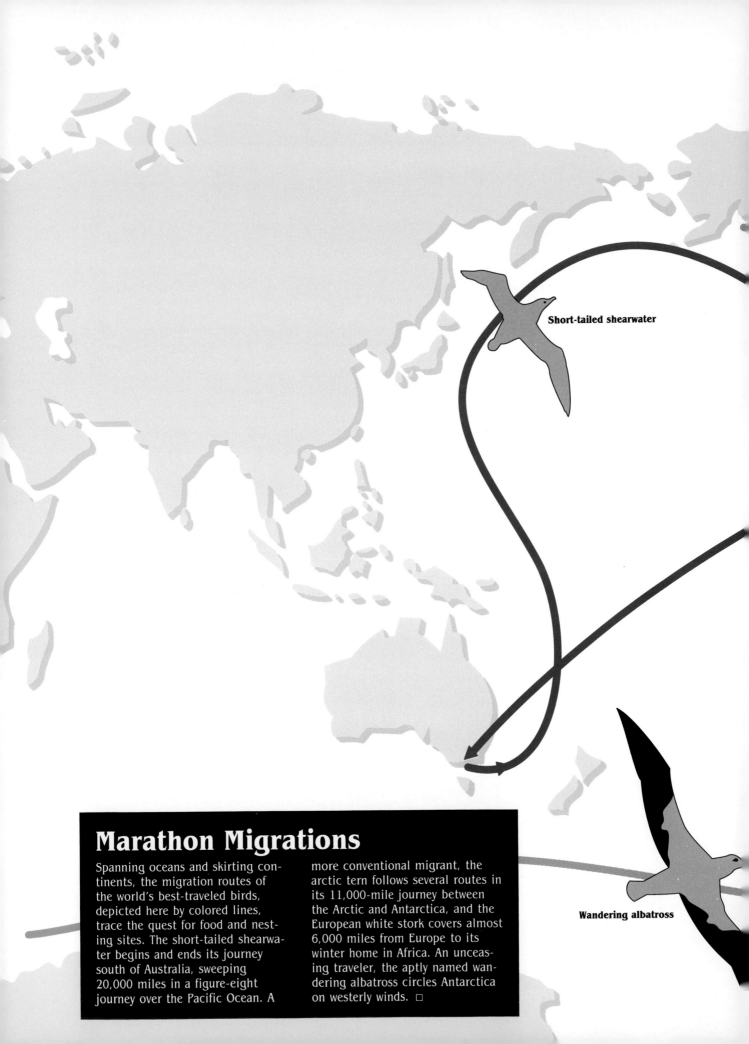

Short-tailed shearwater

Wandering albatross

Marathon Migrations

Spanning oceans and skirting continents, the migration routes of the world's best-traveled birds, depicted here by colored lines, trace the quest for food and nesting sites. The short-tailed shearwater begins and ends its journey south of Australia, sweeping 20,000 miles in a figure-eight journey over the Pacific Ocean. A more conventional migrant, the arctic tern follows several routes in its 11,000-mile journey between the Arctic and Antarctica, and the European white stork covers almost 6,000 miles from Europe to its winter home in Africa. An unceasing traveler, the aptly named wandering albatross circles Antarctica on westerly winds. □

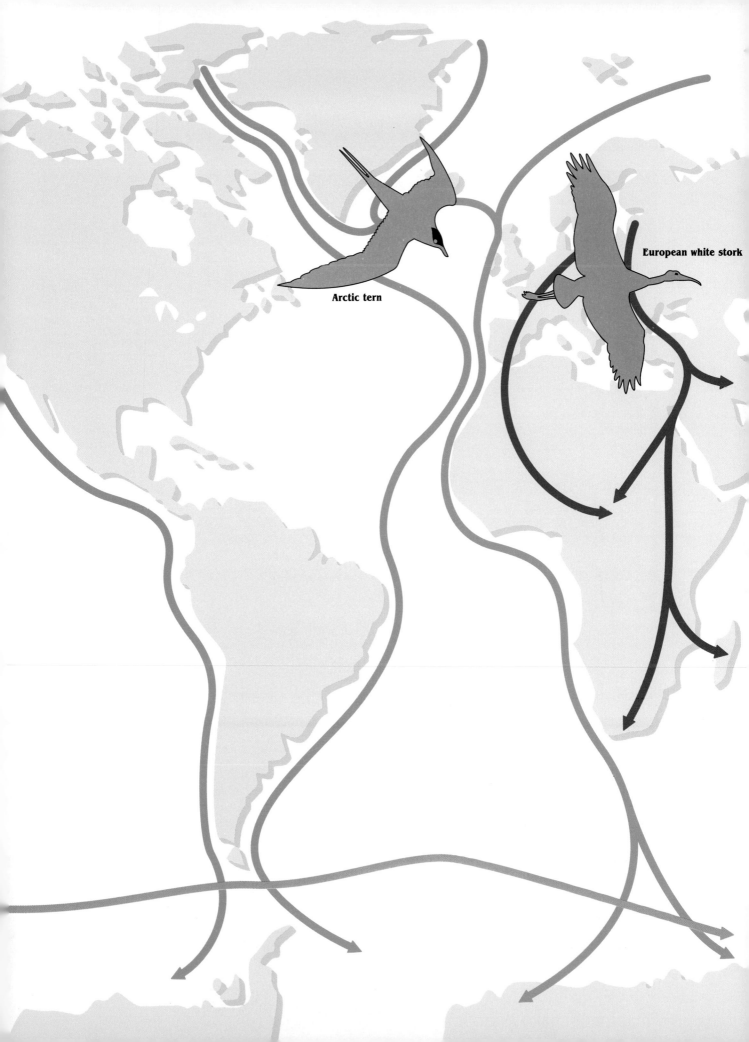

Arctic tern

European white stork

Endless Summer

No animal travels as far or seeks sunlight as persistently as the arctic tern. The tern flies from pole to pole, where the ice-cold waters swarm with rich quantities of crustaceans and other marine life that are staples of the tern's diet. As a result of its migrations, which are the longest of any bird, the arctic tern spends eight months of the year in daylight.

Born in July along the coasts of the northernmost islands of the Arctic Ocean—from Russia and Europe to Greenland and North America—the fledgling tern learns to fly by the time the brief arctic summer begins to wane, its sleek white-and-pearl-grey body developing slender, crescent-shaped wings. By the time that the increased activity of the aurora borealis signals the end of the northern summer with bold streamers of color, the young tern and its family are ready to leave on a migration of more than 11,000 miles to their southern summer home in Antarctica. It is a heady feat for a bird that measures no more than seventeen inches from the end of its curved, blood red beak to the tip of its deeply forked tail.

The tern's migration route is determined by its dependence on a diet of crustaceans. These creatures abound in the near-freezing waters of polar and subpolar regions; they are borne away from the poles by ocean currents. Terns follow these food-rich currents, feeding as they fly—some across the Atlantic Ocean and down the western coast of Europe and Africa, others along the western coast of the Americas.

It is springtime when the terns arrive on the fog-shrouded continent of Antarctica. They have survived the fierce December winds that blow over the Southern Ocean, arriving in a land of towering glaciers and floating ice that they will share with millions of other sea-

birds. Here the terns rest and rejuvenate their exhausted bodies, spending long, bright days molting, feeding, and searching for mates in preparation for the return leg of their migration.

Some terns live to be more than thirty years old, logging in excess of half a million miles. With the world as their flyway, terns regularly make flights of extraordinary length: A tern tagged on the coast of Labrador was captured ninety days later on the coast of southeast Africa, 9,000 miles away. Another, banded in July on the arctic coast of Russia, was recaptured the following May near Fremantle, Western Australia, a record 14,000 miles distant.

Scientists are not certain how terns find their way as they fly. Some speculate that the birds learn the route through years of experience and navigate with the help of the sun and the stars—signposts in the sky that guide the birds to their final destination. □

Precision Flight

Among all the migrating animals, none times its movements with such split-second punctuality as the short-tailed shearwater. Year after year, these sooty-looking, brown seabirds leave the islands of the Bass Strait, south of Victoria, in April. They then journey 20,000 miles around the rim of the Pacific Ocean, returning to their starting point in November. Although the date may vary by a

day—either November 23 or 24—the time of the birds' return is always within the same twenty-minute period of the evening.

The homecoming spectacle builds slowly during the third week of November. Between November 20 and 23, an advance party of several hundred shearwaters glides in and sets about preparing nesting sites for the coming egg-laying season. But this is only the prelude to the fully orchestrated, noisy drama that unfailingly follows. Just at sunset on November 23 or 24, millions of the birds appear on the horizon. Traveling in flocks up to 400,000 strong, they rise from their normal wave-top cruising altitude and circle the islands in clouds so thick that the last rays of sunlight are blotted from the skies.

Then the cacophonous shrieking and beating of wings diminishes as the flocks settle on their nesting grounds; the first birds touch down between 7:40 and 8:00 p.m.—within twenty minutes of their arrival time the previous year. □

With near-perfect timing, a short-tailed shearwater alights on Phillip Island, south of Victoria, Australia, in November 1966.

Antarctic Wanderers

Consigned by nature to roam the skies at the nether end of the earth, the appropriately named wandering albatross *(right)* spends half its life in the dark of the ant- arctic winter and half in the conti- nent's endless summer light.

Its natural medium is the air. On slender wings that span eleven to twelve feet, the wanderer cours- es over the waves in a ceaseless search for food, pressed forward on the relentless westerly winds that circle the polar continent. Rarely does it struggle against the howl- ing gales, choosing instead to ride with them, sometimes soaring for hours without stroking its wings.

These great birds have been tracked on foraging flights of 9,000 miles, searching for the squid and small crustaceans that make up their diet. □

Quitting Capistrano

As long as the oldest residents can remember, the swallows of Goya, Argentina, have flown more than 6,000 miles to nest peacefully under the eaves of the historic Spanish mission of San Juan Capistrano in southern California. Their arrival, often during the week of Saint Joseph's Day (March 19), is so graceful and faithful that it inspired a popular love song of the 1940s—and brought thousands of tourists to the area.

But times change. By the late 1970s, the 300-year-old mission had fallen out of favor with the swallows; most of the annual migrants were passing it by, landing instead at Mission Viejo, 6 miles north, and in other nearby communities. The reason: The town of San Juan Capistrano was no longer a safe place for swallows.

Cliff swallows live in colonies of flask-shaped mud nests, usually tucked under the eaves and in the archways of buildings. The swallows of Capistrano had never restricted themselves to the grounds of the mission itself; any structure with the necessary sheltering overhang would do. For most of the town's history the birds were welcome, but the residents of expensive new subdivisions in Capistrano's hills were less understanding of the birds' sloppy building habits than were the old-timers. Particularly annoying was the swallows' habit of using any handy water source—including swimming pools—to mix mud for their nests. Local law imposes a $500 fine and a jail sentence on those who disturb the birds, but some residents waged psychological warfare by playing loud rock music and hanging

tin cans and toy owls to frighten the would-be nest builders away from nesting sites.

Although the scare tactics have driven away all but a few hundred of the hardiest swallows (at one time they numbered in the tens of thousands), tourists continue to arrive as punctually as the birds: In 1989, more than 15,000 cars were counted entering the town on Saint Joseph's Day. □

A Gift of Crab

Like winged vacationers, waves of shorebirds descend on the beaches on the New Jersey side of Delaware Bay each May. The hundreds of thousands of birds represent eleven species and fill the air with the raucous sound of beating wings and shrieking cries. Having flown up from South America on their summer migration, the travelers are near exhaustion; the 5,000-mile journey consumes most of their reserves of energy. They have just one topic in mind: food.

Simultaneously converging on these same beaches are millions of horseshoe crabs, hardcovered arthropods that have changed little in 300 million years. The crabs migrate to this site for a single purpose: procreation.

The confluence of hungry birds and egg-laying crabs produces a two-week frenzy of feeding and mating, an elemental ritual that takes place annually 40 miles southwest of Atlantic City.

The horseshoe crabs arrive on the shore from the depths of Delaware Bay, the males emerging first as the tide begins to recede. Soon, the females also rise; the creatures mate, and each female lays as many as 80,000 eggs along the high-tide line. Overhead, a million shorebirds circle, waiting and watching as the banquet is spread before them.

It is a rich feast, with impressive restorative powers. The tiny bird known as the sanderling, for example, weighs 1.5 ounces when it arrives on the beach; it will double its weight by the time it leaves. To achieve this, each sanderling may consume 135,000 eggs, and an average flock of 50,000 may eat nearly 7 billion eggs.

Despite the vast quantities of their eggs that are digested each year, the number of horseshoe crabs does not seem to diminish. Each May, they return to provide another feast to the shorebirds on Delaware Bay. □

Migratory birds feast on the eggs of horseshoe crabs at Reeds Beach, New Jersey, during the peak of the crabs' mating season.

Fuel Efficiency

Fat is the fuel of migratory flight. Since migrants rarely stop to eat, the success of their long-distance journeys depends on the amount of energy the birds can store in the form of body fat.

A small bird burns six to eight times as many calories during flight as it does while resting; long flights of migration require so much stored energy that many birds must double their weight before they take off. The bobolink, which flies 5,000 miles from Canada to Brazil and Argentina, bulges with fat before takeoff, giving rise to its nickname, butter-bird.

The blackpoll warbler also doubles its weight before departing on its annual migration from Canada and New England to the West Indies and South America. The bird completes one nonstop, 2,000-mile journey from Canada to Antigua Island in just over three days—an effort that is the metabolic equivalent of a human running a four-minute mile continuously for eighty hours. ☐

A blackpoll warbler, whose weight doubles prior to its annual migration, rests on its perch in Central America.

Dashing Home

As though goaded by the urgency of their mission, migrating birds often journey northward to nest more rapidly than they return to the south. Some European song-

birds make the trip north from central or southern Africa in half the time that it takes for them to fly south to their wintering grounds. Some birds increase their speed markedly as they near their nesting grounds. The blackpoll warbler, for example, averages about 30 miles a day for most of its annual flight from South to North America, but it puts on a burst of speed during the final leg, covering the last 200 miles in little more than a day. □

Graceful migrants, trumpeter swans skim the surface on takeoff from Lonesome Lake in New Brunswick, Canada.

Journey at Sea

Many seals are travelers, gliding gracefully through the sea for several hundred miles between feeding and breeding grounds. But the northern fur seal outdoes them all, with some females migrating 6,000 miles a year to reach their winter feeding grounds.

Each summer, the fur seals give birth and breed on the tiny Pribilof Islands *(below)* in the Bering Sea between mainland Alaska and the Soviet Union. Some 800,000 seals, the largest assembly of one mammal species in a small, localized area, climb the rocky shores. The males arrive first, having completed a short trip of just 500 miles along the coast of Alaska. But the females are the real travelers; by the time they stagger ashore, many have returned from as far south as San Diego, California, where they traveled in search of food-rich waters. Fur seals in the western Pacific make similar long-distance treks, migrating between Tokyo Bay and the Commander Islands in the western Bering Sea and off Sakhalin Island, north of Japan. □

Tiny Titans

Minuscule hummingbirds are titans of flight, small in size but able to propel themselves over mighty distances on yearly migrations.

The smallest bird in the world, the bee hummingbird, carries its bumblebee-size body up to 130 miles across the water from its home in Cuba to Haiti, Jamaica, and Yucatán. The larger ruby-throated hummingbird, which still weighs just a fraction of an ounce, migrates 600 miles across the Gulf of Mexico in a nonstop, twenty-six-hour flight from North to Central America, beating its wings forty times per second—a total of nearly four million wing beats without stopping. □

A male bee hummingbird, shown life-size above, sips nectar from a scarlet bush in southwest Cuba.

Split Personality

The eastern kingbird deserves its Latin name, *Tyrannus tyrannus*, for it is truly a tyrant. Although it is only eight inches long, the bird rules its chosen territory with merciless ferocity. Indeed, kingbirds are such jealous guardians of turf that they have even attacked low-flying aircraft. During the summer, kingbirds range through Canada and the United States, each voraciously snapping up thousands of insects every day. The kingbird's taste is almost indiscriminate: More than 200 varieties of bug are fair game for it. Each autumn, however, just before leaving on the long migration flight south to its winter homes in Bolivia and Peru, the kingbird's nature undergoes a radical change. This lonely defender of territory turns gregarious and joins a flock of fellow kingbirds, with which it flies south. Upon arrival in its wintering grounds, the kingbird scorns its insect diet, becoming a gentle sipper of nectar and fruit juice. What triggers the change of diet and character is not known. But ornithologists say that the effect is to reduce aggression and competition for food with other birds that inhabit the area. □

An eastern kingbird looks out from its perch in Rockport, Texas, where it stalks insects to satisfy its voracious summer appetite.

Vee Formation

Group travel offers many benefits to animal migrants. Some, such as ducks and geese, find that it is most efficient. These long-distance fliers travel in vee formations *(right)*, because each bird is then able to ride the wake of the one ahead—a trick that increases the flock's endurance as much as 71 percent. The leader must work hardest, so each member of the vee takes a turn up front. □

Benign Neglect

Like other birds of its kind, the shining bronze cuckoo of New Zealand *(left)* lays its eggs in the nests of other species. It then embarks on a 4,000-mile migration across the Tasman and Coral seas to New Guinea and the Bismarck Archipelago, leaving its unhatched chicks behind to fend for themselves.

The abandoned cuckoos fare perfectly well; weeks later, having been nurtured by another species, they follow in their parents' flight path, successfully crossing thousands of miles of open ocean without benefit of any sort of instruction. Ornithologists surmise that some inborn guidance system helps them on their way. □

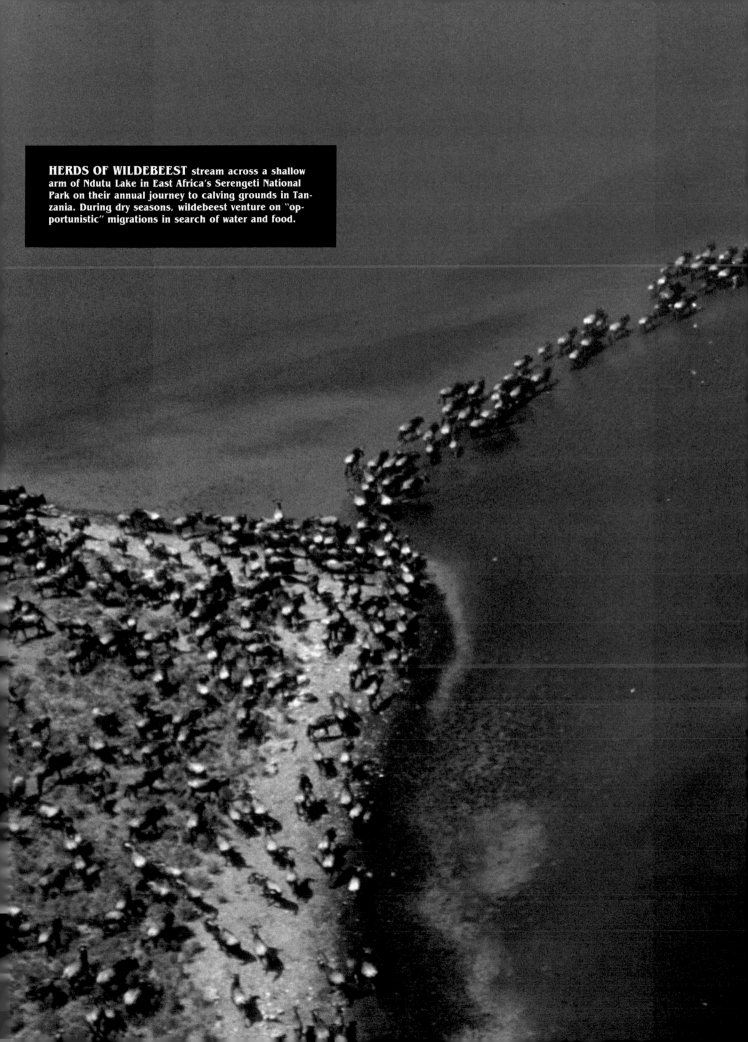

HERDS OF WILDEBEEST stream across a shallow arm of Ndutu Lake in East Africa's Serengeti National Park on their annual journey to calving grounds in Tanzania. During dry seasons, wildebeest venture on "opportunistic" migrations in search of water and food.

A U.S. sailor on Midway Island in 1945 "interviews" one of Midway's albatrosses for the island's radio station KMTH.

Airborne Battle

When the United States Navy decided during World War II to build an airfield on the Pacific island of Midway, northwest of Hawaii, it found itself engaged in battle with a determined airborne enemy—a large flock of Laysan albatrosses. The albatrosses and the navy were attracted to the island for a common reason: As its name implies, Midway's volcanic rocks lie in the middle of the vast Pacific Ocean.

For the navy, Midway was a critical midocean staging point for aircraft attacking Japan. For the albatrosses, the island had always been a favored nesting site. Each autumn, from October to December, thousands of these mariners of the air converge to rest and nest on Midway; to do so, many complete meandering flights that cover some 20,000 miles.

The presence of the birds—occupying runways, hangars, and radio transmitters—made normal flight operations nearly impossible. The albatrosses were particularly attracted to the navy's long runway, which eased their gangling attempts to take flight. (The birds are so awkward on land that they carry the nickname "gooney bird.")

Navy officials first believed they could obtain use of Midway by frightening the albatrosses away with scarecrows. That failed. They tried enticement, placing female decoys on neighboring islands, also to no avail. As a last resort, sailors netted the migrants and hauled them away; some returned from as far as 3,200 miles, and still more appeared, conditioned by generations of migratory habit.

In the end, the birds won the right to stay, and a truce of sorts was made with the navy. An inventive chief petty officer discovered that vehicles driven down the runway with sirens, horns, and other noisemakers blaring would scare the birds aloft and clear a path for arriving and departing airplanes. □

Urban Invaders

During the 1980s, the city of Albuquerque, New Mexico, expanded out of the Rio Grande valley into the rugged foothills of the Sandia Mountains, which loom east of the city. The new residents of the area found spectacular views of the desert, new freedom from urban cares, and new neighbors: a native population of black bears, whose own homes had been disrupted by the development.

Strangers in their own land, the bears had nowhere to go but downtown. They meandered through malls, climbed onto rooftops, even tried out the new suburban residents' swimming pools. Albuquerque's bears are not the only wild animals to find their homeland invaded by humans. Almost invariably, the animals are longtime inhabitants who had come to rely on the ecology of an area to furnish food and living quarters. When people move in, the animals don't know where to go.

Sometimes, however, the animals are the invaders. In 1969, a herd of moose moved into a residential area of Anchorage, Alaska, driven by hunger resulting from a shortage of food in the countryside. Migrating polar bears frequently wander the streets of Churchill, Manitoba, far north on the shores of Hudson Bay. Dazed moose occasionally stroll into towns in Maine, New Hampshire, Vermont, and northern New York; more than once they have caused damage by smashing storefronts or automobiles.

The invasions are not always confined to urban islands in the countryside. Some animals seem to enjoy an urban environment for its own sake. According to a recent census, 2,000 foxes roam the streets of London. And from November to May each year, flocks of snowy owls migrate from northern Alaska, Canada, and Greenland to make their homes amid the shriek and roar of jets from Boston's Logan International Airport. □

Researcher Norman Smith holds aloft one of thirty-five snowy owls that wintered on the runways of Boston's Logan International Airport in 1988.

Super Bees

In 1956, Warwick Estevan Kerr, a honeybee researcher from Sao Paulo, Brazil, made a journey to Africa, where he acquired two dozen of that continent's native wild honeybees that he intended to breed with the European strain he kept in his laboratory. Al-

A WELL-FED BRITISH hedgehog peers out of a sheltering flowerpot. When developers pulled out miles of the country's famous hedgerows to make room for houses, numerous hedgehogs—spiny shrews that lived in the undergrowth and rooted in the ground for food—moved into the new neighborhoods, where families welcomed them as pets, rather than pests, leaving food, water, and shelter for the displaced creatures.

though the domestic European bees are the world's prime commercial honey producers, the African bees are stronger and more industrious. Kerr hoped that by crossing the two strains he might develop a more robust and productive "super bee" that would be better adapted to Brazil's tropical climate.

In one sense the experiment was successful: The two strains, bred together, produced a hardier honeybee than the pure European variety. However, the new bees proved to be militantly defensive and fiercely territorial. Soon, they came to be known by the ominous nickname of "killer bee."

In 1957, twenty-six swarms of the Africanized bees escaped from Kerr's apiary. They swiftly began invading colonies of European bees, mating with their queens and producing still more hybrids. Slowly at first, and then more rapidly, the bees swept across the Americas. They took Venezuela in 1978 and Colombia in 1979. By 1981, they had conquered Central America. By 1990, they had moved across Mexico and were poised to cross the Rio Grande into the United States.

The Africanized bees' thirty-year march has left some 1,000 humans dead, as well as thousands of domestic livestock and pets. But scientists claim that the "killer bee" nickname exaggerates the bees' threat, explaining that the creatures are not wantonly aggressive, just fiercely protective of their colony and queen.

They are indeed fierce. Although the venom of an Africanized bee is no more potent than that of a European one, the tiny terrorists often attack in swarms, producing hundreds of stings in a short time. Five hundred stings are usually fatal, and the only defense is to flee; most bees will give up the chase after a mile or so. □

A wild colony of Africanized bees constructs its hive in a tree in the city of Maturín in northeastern Venezuela.

A swarm of locusts descends on an Ethiopian field in 1968. The year's plague threatened crops in more than forty countries from West Africa to northern India.

Deadly Rampages

Since before the time of Moses and the pharaohs of Egypt, swarms of locusts have swept across the landscape of Africa and the Middle East, darkening the skies and consuming the crops of agricultural civilizations. Equipped with powerful jaws and voracious appetites, locusts—which are, in fact, great masses of migratory grasshoppers—move by the hundreds of millions. Chewing their way across the landscape at more than three miles per hour, locusts have earned a reputation as the most destructive insects in the world. But such claims can be deceiving.

Locusts usually dwell in quiet harmony with the farmers of northern Africa, the Middle East, and India. As grasshoppers, the creatures feed at night on scrubby desert plants; when they become the gregarious locusts, they feed by day, devouring all before them.

The transformation from one to the other takes place rarely, only when weather conditions have promoted runaway population growth. The resulting crowding forces the grasshoppers to congregate in dense clusters. When a swarm reaches a critical density, it goes on the move; the devastating migratory phase has begun. Such a legion of locusts covering one square mile may number between 100 million and 200 million insects. More than a dozen such masses move separately across the land, so that the total infestation may number in the billions.

Temperature and terrain influence the swarm's appearance. Sometimes it moves through the air like mist; sometimes it rolls like the advancing wave of a thundercloud or towers like a tornado. The cloud always moves with the wind, borne toward areas of rainfall and fresh vegetation. Before cooler weather stills its movement, a plague of locusts may travel 2,000 miles in a single season. □

Hunting for Home

In ancient times, results of the Olympic Games were sped throughout Greece by air, written on tiny scrolls attached to the legs of pigeons that had been trained to wing swiftly to their home cities. News of Caesar's conquest of Gaul reached Rome in a similar fashion, and during the nineteenth century, the French military employed homing pigeons to convey information across the battle lines of the Franco-Prussian War.

The birds were also used to good advantage as couriers during World War I, but they are now best known as the poor man's race-horse. Raised in lofts built on urban rooftops, the pigeons are transported far from home and then sent aloft for timed races to their familiar coops.

Homing pigeons are a domesticated variety of the wild rock dove. Though not a migrating bird, the dove has acquired an exceptional ability to find its way home from foraging flights. Through selective breeding, the homing ability has been refined so that trained pigeons frequently cover distances ranging from 500 to 1,000 miles. Scientists have gone to great lengths to discover what it is that provides this and other long-distance animal navigators with their course-finding skills. These researchers have fitted pigeons with frosted contact lenses to block their sight and attached tiny magnets to disrupt any sort of biological compass. The scientists have tested birds' ears and their sense of smell, tracked the birds from small airplanes, and thrown them from moving vehicles—all in an effort to figure out exactly how they find their way home.

The tests have shown that pigeons rely not on a single sense but on many. They can detect ultraviolet light that is invisible to humans, hear low-frequency sounds that are inaudible to humans, distinguish between certain smells, and sense weak magnetic fields. Pigeons also have individual styles of homing; those raised in different places rely on different techniques. This quality has sometimes confused researchers. In 1972, for example, two Italian scientists concluded that smell was a prime navigational tool, because, in their experiments, obstructing a pigeon's sense of smell left it completely disoriented. But when the experiment was tried on German pigeons, the birds had no trouble making their way home. Apparently, they used an entirely different, but unidentified, sense.

So much appears to contribute to a pigeon's navigational network that it is not surprising that the birds themselves occasionally become confused—sometimes spectacularly so. In 1988, 6,000 West German racing pigeons lost their way home from Denmark, a 300-mile trip they should have completed in eight hours. Half the birds eventually turned up, but the rest were gone forever. In another instance, an English racing pigeon released in northern France appeared two months later in Durban, South Africa, some 6,000 miles from its home. □

A homing pigeon wears a pair of frosted lenses to cloud its vision as part of a 1970 experiment to determine how the pigeons navigate.

An Army of Individuals

Every three or four years, an anxiety seizes Norway's lemmings, a tension triggered by a population explosion among the normally solitary, hamster-size rodents. Within days, distress turns to stark panic, and the lemmings begin their famous suicidal trek from their mountain burrows into the fields, roads, and towns of the nation.

Like frightened theater patrons fleeing a fire, the lemmings are a mob of individuals, each one looking out for itself. Although a common terror impels them, the creatures move alone, only coming together rarely, when mountains, valleys, and rivers force them to do so. Then, they sometimes devour those lemmings in their way.

The lemmings' frantic impulses have driven them directly through houses, barnyards, and villages. Whenever they are in their cyclical frenzy, the lemmings die by the thousands from exhaustion, starvation, drowning at sea, or rushing headlong into automobiles—thus producing a carpet of corpses on some Norwegian highways. □

A frantic migrating lemming pauses to bare its teeth at an interfering photographer near the city of Bodø, in northern Norway.

Laurence Mondet of Le Havre, France, recognized her well-traveled cat, Gaston, by the animal's tattooed ear.

Wayward Pigeons

High on Jersey Hill, an undistinguished knob in southwestern New York State, there once stood a fire tower. It was an ordinary enough structure, except for one curious circumstance, discovered in 1967 by Cornell University ornithologists: Homing pigeons released from the tower could not find their way home to Cornell, seventy-five miles away in Ithaca.

The lost pigeons were not mistrained misfits, and they had no difficulty locating their lofts from other release sites outside Ithaca. It seemed that Jersey Hill's fire tower was somehow responsible for confusing the pigeons' navigational sense. But after the tower was torn down in 1988, the mysterious effect continued; apparently, Jersey Hill itself, not the tower, is responsible. To this day, the place confounds Cornell's pigeons as well as their keepers.

Curiously, pigeons raised in places other than Ithaca are not confused by Jersey Hill. The president of the American Racing Pigeon Union once put his prize-winning squadron at risk in order to test the phenomenon; all but one bird effortlessly returned to the loft, located sixty miles away in Rochester, New York. As yet, no amount of experimentation has yielded an answer to the mystery of Jersey Hill. □

Incredible Journeys

When Doug Simpson's pet German shepherd, Nick, wandered away from their campsite in the Arizona desert, the man was convinced that his dog was lost forever.

But four months later, a battered, bruised, and thoroughly exhausted dog—looking and acting, in spite of his injuries, exactly like Nick—limped up to Simpson's car, which was parked in the driveway of his parents' home in Selah, Washington, more than 1,000 miles from the desert where Nick had last been seen.

Thousands of pet dogs and cats disappear from their owners' homes every year, and most are never seen again. But every so often a pet reappears that has apparently tracked its owner over astoundingly great distances.

However, experts in animal behavior reject most tales of long-distance travel. Although an acute sense of smell may enable some animals to find their way home over short distances, scientists doubt lost creatures can find their way over many miles. Researchers speculate instead that grief-stricken owners of lost pets are likely to accept similar-looking strays as their own. But even the skeptics were stumped in 1985, when a sixteen-month-old French cat named Gaston tracked his mistress 320 miles from Le Havre across northern France to the coastal resort town of Marennes. There was no question of the cat's identity: Gaston was recognized by a tattoo that had been etched in his ear when he was a kitten. □

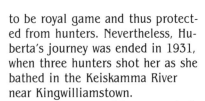

Sonja Zytkow's imaginative memorial to Huberta featured the hippopotamus, her ghost, and a map of her journey.

Huberta the Hippo

Scientists say that hippopotamuses are homebodies, rarely leaving the rivers and muddy wallows of their youth, content to graze on aquatic plants and raise their young in familiar surroundings. They leave the difficulties of migration to the birds and other, more high-strung creatures.

But in 1928, Huberta the hippo, flying in the face of scientific certainty and the traditions of her kind, rose up out of her comfortable lagoon in Zululand, South Africa, and embarked on a three-year, 1,000-mile odyssey that would bring her fame, excitement, and a tragic death.

Huberta—initially dubbed Hubert, until her true sex was discovered—was a celebrity from the moment she was first spied ambling through the sugarcane fields near her home, munching on the long sweet stalks. First she rambled south through the pristine, coastal province of Natal. She was a traffic stopper on railroads and highways. She snacked on gardens and frolicked in the waves near seaside resorts.

She eluded capture by representatives of the Johannesburg Zoo, her every tactic of escape recorded by newsreel photographers. She was once a guest (albeit uninvited) of the Durban Country Club, appropriately emerging on the clubhouse veranda in the early morning hours of April Fools' Day, 1929.

Huberta became a national hero of sorts, and she was proclaimed to be royal game and thus protected from hunters. Nevertheless, Huberta's journey was ended in 1931, when three hunters shot her as she bathed in the Keiskamma River near Kingwilliamstown.

Huberta's story did not conclude with her death. Tens of thousands of people have filed past her stuffed figure, which stands at the entrance to the natural history building of the historic Kaffrarian Museum in Kingwilliamstown. And far away in Sausalito, California, South African sculptress Sonja Zytkow erected a memorial to Huberta in the mid-1970s. Zytkow's monument was destroyed, however, in the 1989 earthquake that rocked northern California. □

Grasshopper

Deep in the remote heart of the Beartooth Mountains of Montana lies Grasshopper Glacier, a giant field of ice shot through with dark stripes that bear mute testimony to the capriciousness of nature. The glacier takes its name and color from the bodies of billions of Rocky Mountain locusts buried in it. Through thousands of years of migrations, millions of locusts during each passage were stunned by the icy mountain air, fell to the glacier's surface, and died. The insect carcasses, covered by successive layers of snow, became embedded in a window of ice.

The last recorded locust flight through the Beartooth Mountains took place in 1898. □

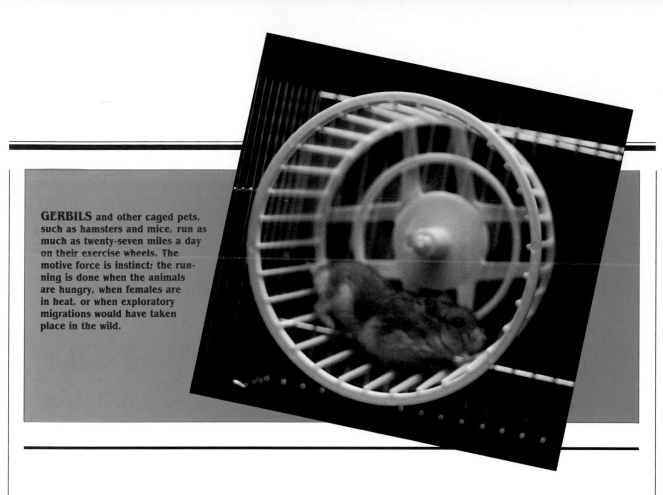

GERBILS and other caged pets, such as hamsters and mice, run as much as twenty-seven miles a day on their exercise wheels. The motive force is instinct; the running is done when the animals are hungry, when females are in heat, or when exploratory migrations would have taken place in the wild.

Icy Passages

It was once believed that polar bears were ice-borne argonauts of the animal kingdom, traveling purposively through their frigid world on ice floes propelled by ocean currents.

Modern research has found that the bears instead trust to luck for their navigation: When they hitch rides on ice floes, as often as not they find themselves floating off in random directions *(left)*. But such meanderings do not necessarily lead to hardship. While afloat, the bears are quite likely to encounter other migrating wildlife, such as seals and walruses, which they eagerly attack for food. And when their frozen transports near the shore, the polar bears simply debark and continue their journeys across land. □

Mule deer descend the snowy, steep terrain of the Rocky Mountains near Almont, Colorado.

Vertical Migrations

Most animals migrate horizontally between their summer and winter homes—usually north to south. A few, though, demonstrate that a vertical migration works just as well, for a relatively small change of elevation can produce a dramatic change of climate.

Ladybugs, mule deer, moose, elk, and many species of birds move up into the cool mountains for the summer and spend their winters in the warmer valleys. The mountain chickadee of the Rocky Mountains and the Sierra Nevada migrates a mere 4,000 vertical feet each spring and fall, but it achieves a change in climate equivalent to flying 1,200 miles north or south. □

THE ANIMAL MIND

There was a time when grieving gorillas, tipsy elephants, and birds that speak in numerous dialects would have seemed beyond possibility. Animals were not deemed capable of reasoned thought, emotions, and humanlike behavior. Their minds—if they had minds in any real sense—were thought to be sufficient to handle instinctive actions, but more complex mental processes were beyond the range of the animal kingdom.

This comfortable notion has been shattered by modern research, and it now appears that animals have far greater mental abilities than previously suspected. Perhaps more intriguingly, they seem to experience a wide range of what we call emotion. Such discoveries indicate that there are unplumbed depths of the animal mind. The animal kingdom still holds many secrets—and humankind may have much to learn from the creatures that share its world.

"Fine Animal Gorilla"

It was a scene from some absurd, fanciful nursery: Koko, a 260-pound gorilla, dwarfing her visitors, was being asked to show off by playing a simple baby's game. Koko's enthusiastic trainer at the Gorilla Foundation near San Francisco, Francine Patterson, pointed to her own eye, expecting Koko to follow suit, as she had many times before. Instead, the gorilla pointed to her own ear. Patterson tried again, this time pointing to her nose. Koko touched her chin. Annoyed and frustrated, Patterson scolded the gorilla. In the American Sign Language—widely used by the deaf—that she had taught her giant friend, Patterson spelled out a stern reproach: "Bad gorilla." Deadpan, Koko demurred, using her own hands to spell out her reply. "*Funny* gorilla," she corrected her reproacher.

Koko was indeed funny. But she soon proved to be an extraordinary creature in other ways, too. Born at the San Francisco Zoo in 1971, she first used American Sign Language as a one-year-old, learning to associate symbols with objects just as a hearing-impaired human would. Patterson, then a graduate student, taught Koko the signs by forming the gorilla's hands into the proper shapes. Eventually, according to Patterson, Koko would learn some 500 sign language words. She learned to fling insults—"you dirty bad toilet" was a favorite—at her trainers and even learned to fib in the manner of a naughty child, once unsuccessfully attempting to blame a broken sink on her trainer.

Like any child struggling with language, Koko often seemed to invent words when she did not know an object's proper name. A cigarette lighter, for example, became a "bottle match," and a ring was a "finger bracelet." Sign language, like any other language, is susceptible to personal and regional accents, and Koko developed what Patterson calls "a heavy gorilla accent," touching her body to express herself far more than a human signer would.

To be sure, some scientists dispute such accounts of Koko's performances, contending that they may be based as much on Patterson's well-meaning wishful thinking as on the gorilla's true language skills. But Patterson insists that Koko learned to express herself in some detail, telling about apparent emotions and discussing such matters as aging and illness. One such dialogue between the gorilla and Patterson reportedly went as follows:

"Where do gorillas go when they die?" Patterson asked. "Comfortable hole bye," came Koko's signed response.

"When do gorillas die?"

Huge but gentle, Koko the gorilla cuddles her pet kitten Red.

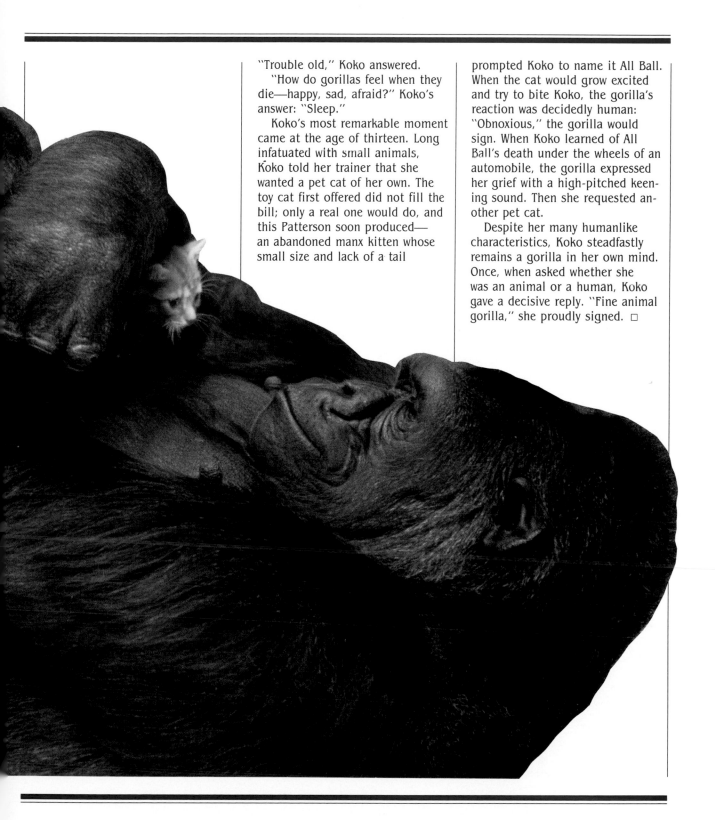

"Trouble old," Koko answered.

"How do gorillas feel when they die—happy, sad, afraid?" Koko's answer: "Sleep."

Koko's most remarkable moment came at the age of thirteen. Long infatuated with small animals, Koko told her trainer that she wanted a pet cat of her own. The toy cat first offered did not fill the bill; only a real one would do, and this Patterson soon produced—an abandoned manx kitten whose small size and lack of a tail prompted Koko to name it All Ball. When the cat would grow excited and try to bite Koko, the gorilla's reaction was decidedly human: "Obnoxious," the gorilla would sign. When Koko learned of All Ball's death under the wheels of an automobile, the gorilla expressed her grief with a high-pitched keening sound. Then she requested another pet cat.

Despite her many humanlike characteristics, Koko steadfastly remains a gorilla in her own mind. Once, when asked whether she was an animal or a human, Koko gave a decisive reply. "Fine animal gorilla," she proudly signed. □

Handy Orangutans

While his keepers' backs were turned one day in the early 1900s, a crafty orangutan by the name of Dohong decided to make a break from New York City's Bronx Zoo. This animal, like many other orangutans, had a particular skill with tools. So when Dohong saw his chance, he seized a piece of metal from his trapeze and deftly pried apart the wire mesh that separated him from the outside world and his freedom.

In captivity and in the wild, orangutans often press into use as makeshift tools whatever materials they discover at hand. Among human beings, whose presence makes a wider range of materials available, the animals can be quite creative. In the wild, wooden sticks are especially popular; orangutans will use the sticks for scratching, to extend their reach, to drive off bothersome wasps, or to pierce coconut shells. Orangutans have even displayed a certain nautical flair, catching laborsaving rides to distant feeding sites on floating pieces of driftwood. □

A young orangutan in tropical Borneo employs a leaf to shield itself from the harsh sun.

Imo, the Genius Monkey

Just as the old saw says, a monkey seeing is usually a monkey doing. However, even though monkeys make good copycats, there are some, like old dogs, that do not readily take up new tricks.

In 1952, biologists on Koshima Island, in southern Japan, began scattering sweet potatoes on the sandy beach near their laboratory, hoping to observe the food-gathering behavior of a troop of macaque monkeys living in a nearby forest. The monkeys accepted the food, although they had difficulty brushing off the coating of sand the potatoes acquired. Then one day, a young macaque the scientists named Imo contrived a way to overcome the distasteful inconvenience: Imo carried her sand-caked potato to the sea, where she washed and scrubbed it with her hands. Soon, a playmate imitated her behavior; Imo's mother followed suit, and before long, most of the troop were washing their sweet potatoes.

Imo's reputation as an innovator was enhanced in 1955, when the Koshima Island biologists began scattering wheat on the beach. If sandy sweet potatoes had been an inconvenience to the monkeys, wheat mixed with sand presented them with a nearly insurmountable obstacle; the wheat could be eaten only after painstakingly picking out individual grains from handfuls of sand particles.

However, Imo soon solved the problem by simply tossing handfuls of the sand and wheat mixture into the sea. The sand sank, while the clean wheat floated to the top, ready to be skimmed off and eaten. Once again, Imo's discovery spread through her troop.

But some never caught on. Most unteachable were older males, who have little contact with younger animals from whom they might learn new tricks—even those as simple and beneficial as Imo's. □

Following a practice originated by the inventive Imo in 1952, a Japanese macaque washes sand from a potato.

Partners in Peril

A troop of baboons will sometimes rely on the keen senses of ungulates, such as gazelles and impalas, to provide an early warning of a predator's approach. To an extent, these species understand each other's languages—or at least their cries of alarm. The sound of a gazelle's warning bark is enough to galvanize baboons into instant defensive action.

The frailer ungulates also benefit from the arrangement. Not only do they heed the baboons' warning calls, they have also been known to rely on the primates' superior brawn. One herd of impalas was observed feeding alongside a baboon troop when a trio of cheetahs appeared in the distance. The impalas, though clearly alarmed, resisted their natural impulse to take flight, which would have allowed the swift-running cheetahs to separate one or two from the herd and kill them. Instead, the impalas stood their ground as the baboons (whose strength and daggerlike canine teeth make them more than a match for most predators) drove the cheetahs off—just as the impalas had clearly expected the baboons would. □

Two male baboons, often competitors, join forces to chase away an interloper.

Battling Baboons

When courting a female in heat, a male baboon will sometimes call on a friend for help—not in over-powering the female, but in driving off competitors.

The suitor baboon signals his need for help by turning his head from side to side, even as he stares down his rival. When help arrives, the pair frighten off the lone male competitor, allowing romance to take its natural course in favor of the original beau.

What is remarkable about this buddy system is that the backup baboon—who answers the call for help—risks injury but receives no immediate reward for his assistance. Zoologists find that the favor—if it is that—frequently is repaid at some time in the future; in most cases, when the roles of the two baboons are reversed, the one who earlier won a mate will rush to the aid of his erstwhile rescuer—certainly a fitting payment on the IOU.

On the other hand, a beneficiary who does not return the courtesy is punished: In future encounters, he will be left to handle an opponent on his own. □

The Enigma of the Talking Chimp

Although the structure of his vocal apparatus prevents him from speaking, a pygmy chimpanzee by the name of Kanzi has proved to be a talented linguistic do-it-yourselfer: Kanzi appears to have learned to understand speech the way human children learn, by watching and listening.

A resident of the Language Research Center in Atlanta, Georgia, which studies language abilities of primates, Kanzi has exhibited an unprecedented command of human language. For example, if the chimpanzee is told, "I hid a surprise under my left foot," the animal will unhesitatingly rush over and lift his attendant's left foot. Kanzi's abilities are real: Scientists test the chimpanzee by playing recorded phrases, which Kanzi hears through headphones, so that his keepers cannot give him clues—even inadvertently—to what is being said.

What is more, Kanzi appears to have acquired his understanding spontaneously, without any rigorous training, making him the first animal known to have developed language skills in a manner similar to that of humans.

Kanzi has been taught to speak to his trainers using a computer keyboard marked with geometric symbols, each one representing a different word. By combining the symbols, Kanzi is able to form rudimentary phrases and to make simple requests.

Headphones in place, Kanzi listens to a researcher's recorded instructions. Recordings prevent scientists from inadvertently giving the chimp subtle clues that might guide his actions.

In the course of a typical day, Kanzi might ask to be tickled, try to get up a game of tag, or ask someone to accompany him to his tree house for a banana. Occasionally, the chimpanzee will ask to watch television; his favorite show is a videotape of the primatologist Jane Goodall living among the chimpanzees in Africa.

Kanzi's accomplishments are teaching scientists much about the way animals communicate. But researchers also hope to unlock secrets of human evolution, for Kanzi's achievements suggest to some that humankind's distant ancestors may have had more advanced language skills than previously suspected. □

Primate Grief

Do animals mourn their dead? Humans tend to think of the so-called higher emotions—such as grief—as being uniquely human in character, beyond the experience of the animal world. This view, however, may reflect human self-centeredness more than reality, for time and again, animals have demonstrated behavior that appears to express profound grief.

Gorillas seem to be especially given to displays of sorrow. Baby gorillas who lose their mothers will frequently show symptoms of depression that are familiar to every human being, complete with a loss of appetite and general listlessness. This condition may last for as long as a year. Mother gorillas, conversely, will sometimes take comfort in childlike behavior after the death of an infant.

The well-known naturalist Dian Fossey, who studied mountain gorillas by living with them in the wilds of Zaire and Rwanda for nineteen years, reported that after the death of one gorilla, its comrades milled about for several days as if looking for someone, alert to any random noise that might have been a gorilla call.

Similarly, primatologist Jane Goodall wrote of the brother of a deceased wild chimpanzee that she had named Gregor. For a period of nearly six months, the brother repeatedly returned to the place where Gregor had spent the last days of his life. There the brother "would sit up one tree or another, staring around, waiting, listening."

Although grief is clearly their reaction to death, the animals' response to sickness is more difficult to characterize. In one incident, as an elderly female gorilla lay close to death, her companions flew into a hysterical frenzy, during which they pounded the sick gorilla with their fists, pulled on her limbs, and even jumped on her prostrate form.

Some scientists say the gorillas' abusive actions may have been intended to encourage signs of life, just as humans might shake or slap an unconscious person. Others, though, speculate that the violent gorillas were somehow enraged at their stricken companion and were attempting to deliver a deathblow. ☐

A grieving gorilla mother still clings to the body of her infant the day after its death.

The Grieving Elephant

In Tanzania's Serengeti National Park in 1970, ecologist Harvey Croze witnessed a remarkable natural drama. Croze and a friend had been following a herd of elephants when they noticed an old and sickly cow, probably the matriarch of the group, lagging a few steps behind. Without warning, the elephant sagged to her haunches. Then, with a fierce convulsion, the creature crashed onto her side. Immediately, the other elephants in the herd formed a protective circle around their fallen companion. Each tried in turn to aid her. Several young bulls placed their trunks into the fallen elephant's mouth, trying repeatedly to push her back up onto her feet. Another attempted to nudge the dying creature by placing a foot on her back, as if to rouse her from sleep.

The efforts continued throughout the afternoon, but to no avail, and at last the elephant died. For several hours, the surviving members of the herd remained at her side, as if in deep mourning. One old bull, acting out of either grief or frustration, attempted to mount the fallen elephant in a last frantic attempt to revive her. Finally, the herd moved off, leaving one female to maintain a lonely vigil over the corpse. At nightfall, she, too, slowly withdrew.

These displays of apparent grief are common among elephants.

Although the "elephant graveyard" appears to be a myth—the presumably sacred ground where the creatures go to die has never been found—elephants are commonly observed displaying almost-human signs of grief and mourning.

Some elephants, for example, will attempt to cover the body of a dead companion with leaves or grass. Others have been known to show a fascination with the skeletal remains of fallen comrades. Although they generally ignore the bones of other animals, an elephant skeleton prompts a systematic examination, with special attention paid to the head and tusks, as if in an effort to recognize the deceased. On occasion, elephants have actually removed the tusks from their dead companions as though observing some form of ritual. In a few cases, the tusks have even been smashed against trees and rocks.

Whether this curious behavior constitutes actual grief remains a subject of scientific debate. Those who support that explanation point to the actions of a mother elephant whose calf had died. She carried the corpse of her newborn baby on her tusks for days after its death. Long after the infant's body began to decompose, she refused to relinquish it—activity that might well be evidence of a mother's anguish. □

A victim of starvation in Zimbabwe, an emaciated young elephant approaches the body of a dead companion.

The Elephants' Silent Language

For years, hunters and naturalists speculated that elephants must possess an almost magical means of communication, perhaps a kind of extra-sensory perception that is beyond the range of normal senses.

How could it be, they asked, that as many as 100 elephants would suddenly freeze in their tracks as if on cue, or change direction as if commanded, without any audible warning or signal? What mysterious link allowed scattered groups of elephants to coordinate their movements, even though they were separated by miles of dense jungle, allowing them to converge on a water hole almost simultaneously?

More recently, scientists have discovered that elephants do in fact possess a language outside the scope of human senses—one less mysterious, but more beguiling, than previously thought.

While visiting a zoo in Portland, Oregon, in 1984, Katharine Payne, a researcher at Cornell University, noticed a strange vibration as she stood close to the zoo's elephant cage. It was "a palpable throbbing in the air, like distant thunder," she recalled. The source, Payne

found, was the elephants themselves, whose cavernous body structures allow them not only to trumpet and bellow, but to make other noises of such low frequency that humans cannot hear them. These sounds, called infrasounds, were the cause of the vibra-

tions that Payne felt, rather than heard, in the Portland zoo.

After spending several months recording and documenting the calls of the elephants with sophisticated electronic equipment, Payne discovered more than 400 distinct elephant calls—two-thirds of them infrasound that the elephants and the instruments detect-

Its great ears at attention, an elephant listens to the calls of others across the East African savanna.

ed, but which Payne and her assistants never heard.

Only a wrinkling and fluttering of the elephant's forehead betrays that the animal is communicating. Like the rumblings of an earthquake, which it resembles, infrasound can travel great distances, making it critical to the elephants' mating habits. The females are receptive for only a brief time—perhaps as little as two or three days every four years. By emitting a series of infrasonic mating calls, a female summons agreeable males from miles around.

By observing elephants and recording their calls in the wild, naturalists have managed to "decode" this and other low-frequency signals—including a "greeting rumble," exchanged among elephants who encounter each other after a separation; a "let's go," signaling that it is time for the herd to move on; and a "contact call"—punctuated by periods of ear waving and silence as the animals listen for a response—used to locate other elephants at a distance. □

Boozy Elephants

Elephants, renowned for their long memories, may resort to alcohol to forget their troubles. According to some researchers, the animals can react to stress in a peculiarly human fashion—they get drunk, when they get the chance.

Wild elephants sometimes become inebriated by feasting on rotted fruit, whose fermentation causes it to contain alcohol. Thus under the influence, the elephants bellow loudly, flap their ears, flail their trunks, and must lean against whatever is handy to keep from falling down. Those who do not find a support have been seen to topple to the ground, unable to rise again until the haze of alcohol has lifted. After gorging on overripe mgongo, a plumlike fruit, African elephants have been known to wreck whole villages with their boozy behavior.

While some incidents of elephantine drunkenness are no doubt accidental—the creatures simply stumble upon fermented fruits—others may be quite deliberate. Ronald Siegel, a psychopharmacologist at the University of California, Los Angeles, has reported that when elephants are under stress—forced to live in small spaces with a limited food supply, for example—they are prone to seek solace in alcohol. When allowed to roam free, however, the animals are less likely to do so. □

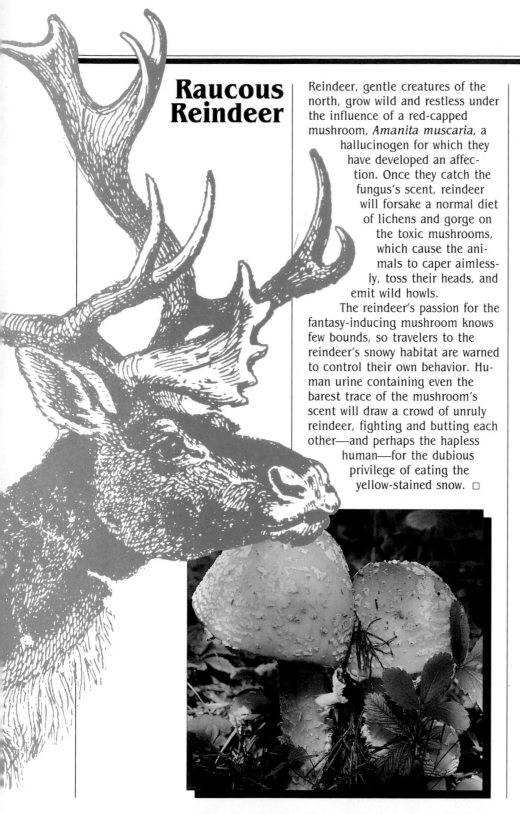

Raucous Reindeer

Reindeer, gentle creatures of the north, grow wild and restless under the influence of a red-capped mushroom, *Amanita muscaria,* a hallucinogen for which they have developed an affection. Once they catch the fungus's scent, reindeer will forsake a normal diet of lichens and gorge on the toxic mushrooms, which cause the animals to caper aimlessly, toss their heads, and emit wild howls.

The reindeer's passion for the fantasy-inducing mushroom knows few bounds, so travelers to the reindeer's snowy habitat are warned to control their own behavior. Human urine containing even the barest trace of the mushroom's scent will draw a crowd of unruly reindeer, fighting and butting each other—and perhaps the hapless human—for the dubious privilege of eating the yellow-stained snow. □

Catnip Craze

Cats go wild over catnip; one sniff can trigger a feline frenzy of shaking, rolling, and leaping. The reason is simple: Ordinary catnip contains a powerful hallucinogen and sexual stimulant that many cats are unable to resist.

In its natural form, catnip is a perennial herb, *Nepeta cataria,* with downy leaves and a strong odor of mint. The herb grows in such diverse locations as Kashmir, Scandinavia, and New Jersey. The hallucinogen it contains is called nepetalactone, and it is remarkably similar to the natural hormones produced by male cats. Some felines, however, lack a certain gene that triggers the reaction and remain blissfully beyond the scent's influence, turning up their noses and walking away when it is offered to them. □

Plumb Loco

Numbers of horses and cattle in the American southwest suffer from the disease of drug addiction: They are hooked on locoweed, a flowering plant that can send even the most imperturbable animal into a crazed mania.

The aptly named locoweed—its name includes the Spanish word for "crazy"—contains a natural alkaloid that, when ingested, can cause powerful neurological and physiological effects. A weed-crazed horse might charge blindly through a barbed-wire fence or lash out at empty space as though battling a snake or a wolf. The plant is addictive, and, like other addicts, animals in its grip will forgo all other food and water for locoweed, at times even starving to death. □

The Language of Fear

A ground squirrel responds to a threat by issuing a distinctive alarm call.

The anxious chatter of a monkey or the squeal of a squirrel that sights a cat are more than simply cries of alarm. Indeed, they are part of an often elaborate, precise language that can mean the difference between life and death.

Scientists have learned that a number of creatures have a far wider vocal repertoire than was once believed. The ground squirrel, for example, gives a sharp squeal when it is in danger; if a predator is pursuing it, the squirrel adds a short trill to the end. Still more embellishments are added to identify its attacker so that others can take appropriate evasive action. If the enemy is a hawk, for instance, the other squirrels are safe in any burrow. But in order to escape a badger, the squirrels must find a burrow with a back door for use as an escape hatch.

Vervet monkeys, whose enemies include eagles, leopards, and pythons, have also developed a complex system of specific alarm. When one of the monkeys spots a leopard, it immediately makes a raspy barking sound that sends any vervet within earshot scampering high into the fragile limbs of tall trees, where even agile leopards cannot follow. A short grunt signals danger from a circling eagle, setting the monkeys scurrying for the protection of dense bushes. A third, high-pitched, warning call alerts the creatures to danger from an approaching snake. Upon hearing this call, the vervets will rise up on their hind limbs and scan the ground, ready to club the snake with their fists. □

The Wonder Horse

Despite astounding displays of humanlike intelligence by many animals, most researchers maintain a healthy measure of skepticism about such accomplishments. All too often, they warn, there are other explanations for supposed shows of animal brainpower.

A favorite example is the case of Clever Hans, the so-called wonder horse of Berlin, who gained renown in 1904 for what seemed to be almost-human intelligence. When his owner, an elderly mathematician named Wilhelm von Osten, wrote a simple arithmetic problem on a chalkboard, Hans the horse would stamp out the answer with his hoof. In answer to the question, "4+5=?" Hans would tap his hoof nine times. Using a code in which letters were substituted for numbers, Hans could also spell out simple words.

Von Osten was fiercely proud of the horse's achievements. He frequently invited scientists to his farm to witness demonstrations, and he seemed eager to have them discover the source of his horse's strange gifts. Finally, two professors from the University of Berlin found that the horse could answer questions only when his owner stood in plain sight. By observing von Osten's actions, the professors soon arrived at a solution to the mystery: Whenever Hans began stamping his foot, von Osten grew rigid and anxious, even holding his breath as he awaited the outcome. When the horse reached the correct number of foot stamps, von Osten would relax slightly and breathe normally.

Incredibly enough, von Osten was not deliberately cueing Hans; instead, the horse himself had learned to respond to this slight change in his master's bearing. Although von Osten's body moved less than one-quarter inch, Hans understood that it was time to cease counting. □

Professor Wilhelm von Osten and Clever Hans pose with an abacus and blackboards, props for the horse's supposed calculations.

Blood Brothers

The fearsome appearance of the vampire bat, the bloodsucking predator of the night, may actually conceal a kindly heart and altruistic nature—at least so far as its own kind is concerned. Scientists have learned that vampire bats will literally form a bond of blood with one another—sharing their meals of blood in order to save the lives of roost mates.

Each night, swarms of vampire bats leave their roosts and look for warm-blooded prey, usually horses and cows. Once a bat lands on a victim, its heat-sensitive nose enables it to locate a patch where the blood vessels are near the surface. Piercing the hide with razor-sharp fangs, the bat will spend as long as one-half hour lapping up the blood of its host.

Vampire bats must consume anywhere from 50 to 100 percent of their body weight in blood each night in order to survive—not always an easy achievement—and on any given night, perhaps one-third of all the bats in a roost fail to gain a blood meal. A bat that goes without food for two nights will die, unless rescued by a roost mate willing to share.

The process by which bats share their blood is surprisingly tender. A hungry bat will show its need by licking the wings and lips of its potential savior in a manner similar to the mating rituals of other species. If the blood-carrying bat is in a generous mood, it will regurgitate a quantity of blood into its mouth. The hungry bat will then feed directly from the mouth of the donor.

In all likelihood, one night's recipient bat will return the favor on another night; a roost of vampire bats quickly forms a network of "blood brothers"—a food-sharing buddy system whose mutual aid will ensure the survival of all. Often, a pair of bats will become life partners in the most literal sense, forming a reciprocal bond to exchange blood on a more or less equal basis. Frequently, but not always, the partners are related to each other.

Food sharing of this sort is rare in the animal kingdom. Only a few other species—including hyenas, chimpanzees, and wild dogs—exhibit a similar generosity, and some zoologists believe that without it the vampire bat might well have died out long ago. □

Bird Language Barriers

Bird fanciers have long delighted in the pleasing melodies and natural elegance of birdsongs. And there is much to delight the ear in this subtle and expressive means of animal communication.

Sometimes, however, understanding falters. Birds of a feather may share the same language, but when a species is numerous and widely scattered geographically, its song develops regional dialects so pronounced that members of the same species can no longer understand one another.

An English chaffinch, for example, cannot communicate with its cousins from Finland or those from central Europe. Willow tits living in the mountains speak an entirely different language from their lowland relatives. And herring gulls from the state of Maine are unintelligible to their relatives in Holland and France. □

These European gulls utter calls in a dialect incomprehensible to American gulls of the same species.

Talking Eggs

Just as human babies will turn and kick within the womb, many birds make a variety of sounds while encased in their eggs. But the cheeping of unborn birds often seems to have a definite purpose.

Chicken embryos emit sounds that are actually calls to their mother to warm her eggs or cool them. In addition, when the mother turns her eggs, the unborn chicks sometimes produce a contented cheeping call. Evidently, the eggs hear as well as speak—all noises stop when the mother gives even the softest cry of alarm.

Eggs that are almost ready to hatch can become downright talkative. Blackbird eggs, for example, produce a sharp, regular clicking sound, which grows more frequent as the eggs approach hatching. Among the most talkative eggs, it seems, are those of mallard ducks. A clutch of ten or so eggs, laid over a period of as many as twelve days, appears to communicate in order to ensure that the eggs all hatch within a few hours of each other. Clutches of eggs kept together hatch almost simultaneously, while those held separately hatch at random intervals over a period of two days. Evidently, more than a mere preference for synchronicity causes the siblings to keep in touch. Within hours of the first birth, mother and offspring are on the move, abandoning any chicks not yet hatched. □

Talented Birds

In 1989, a myna bird named Simon delighted St. Petersburg, Florida, radio listeners with his tuneful rendition of "Blueberry Hill," Fats Domino's rock-and-roll classic of the 1950s. Simon's version of the well-known song, recorded at his domicile in nearby Pinellas Park, became a minor sensation when it hit the airwaves, drawing more requests than such popular human entertainers as Elvis Presley and the Beatles.

Simon may be the first myna to use its natural gift for mimicry to achieve radio stardom, but this talent is by no means unique. Many species of birds have an innate talent for imitating not only human voices, but a variety of other sounds as well.

The European marsh warbler can mimic as many as seventy-five other birds. The aptly named and vocally prolific North American mockingbird, whose Latin name, *Mimus polyglottos*, means "many-tongued mimic," has been known to imitate fifty-five different species of birds during a single hour. Its repertoire can also include mechanical sounds, such as the creaking of a door or the squeak of a wheelbarrow. Bowerbirds also have great range, producing sounds from a cat's meow and a dog's bark to an automobile horn and the crisp thunk of a woodcutter's ax.

Although the purpose of these wide-ranging, curious talents is debated, ornithologists speculate that mimicry serves the same territorial function as any other bird call, announcing a bird's claim on its domain. In captivity, the imitation of human voices may be a reflection of a bird's attachment to its keeper. Parrots, for instance, are more likely to talk when their owners leave their sight, apparently

Versatile mimics, parrots are among several types of birds that are capable of imitating human voices and the calls of other species.

in an attempt to attract the people back into the room.

Sometimes, however, a talking bird will become resolutely silent, often at inopportune moments. An official at the National Zoo in Washington, D.C., once trained a myna bird to repeat the phrase "How about the appropriation?" hoping to impress Calvin Coolidge during a scheduled presidential visit. But when the president arrived, the myna bird refused to utter a single word.

Ironically, another bird was exiled from the zoo a few years later for speaking up: His vocabulary was ruled too racy. □

Milk Battle

Early one morning in 1921, a housewife in Swaythling, England, opened her front door to bring in the bottle of milk that had been delivered before dawn. To her surprise, she discovered that she had been robbed—the bottle's cardboard stopper was missing, and an inch or so of milk had been drunk.

She was not alone; several neighbors reported the same mysterious theft of milk. And when the milkman was questioned, he offered a possible explanation: Every morning a flock of eight or nine blue tits—a tiny blue and yellow bird no larger than a walnut—followed him along his delivery route. Perhaps the birds were the culprits, he suggested.

Indeed, they were guilty. The next morning, a victim observed the birds' modus operandi: No sooner would the milkman leave his delivery than a blue tit would flutter down from the nearby trees. Gripping the rim of the milk bottle with its feet, the bird repeatedly hammered the cardboard stopper with its beak until the top popped open, then carefully removed it and disposed of it in a nearby bush. After throwing away the top, the bird returned to help itself to a hard-earned drink of milk.

Before long, families all over England were complaining of stolen milk, and within a few years, the epidemic had extended into Scotland, Ireland, and Wales.

Evidently, the behavior had spread through the blue tit population by imitation, with each bird passing the information along to its peers.

By 1944, the problem had become so widespread that Britain's milk industry switched from cardboard to metal stoppers. Still the milk thievery continued. Customers were then instructed to leave stones, pots, or towels outside their homes at night, so the milkman could cover the lids of the bottles as he delivered them.

Again, the blue tits rallied. One man watched in fascination as a bird energetically bounced up and down on a stone covering his milk bottle, causing it to teeter and fall. Another awed patron witnessed her towel flying through the air, borne aloft by a team of blue tits.

For the nature-loving British, the wave of milk stealing posed a genuine dilemma: No one wished to harm the birds, yet the thefts were a considerable nuisance. But Harry Champion, dean of the Oxford Forestry School, stood ready with one tongue-in-cheek solution: "I guess we'll simply have to abandon the [British] Isles," the dean proposed. "We can't change our habit of putting milk on our stoops, and we can't harm the birds."

In the end, British householders muddled through. They still have the bulk of their milk delivered to their doorsteps and have learned to live with the occasional depredations of thieving birds. □

A diminutive British thief, this blue tit opens the top of a milk bottle in order to steal a drink.

Dance of the Robot Bee

In August of 1988, a decidedly odd-looking bee entered a hive in a German laboratory. Smeared in wax and doused with the scent of bees, this bee had no sting, for it was a robot, a computer-operated simulation of a bee.

Once inside the hive, the robot began a series of programmed movements, turning this way and that, buzzing, and doling out drops of sweetened syrup—all intended to tell the real bees that food had been discovered almost a mile away, to the southwest. To the relief of the scientists, scores of actual bees responded instantly, flying directly to the spot indicated. In a very real sense, humans had learned to talk to animals.

Once, it was widely assumed that bees conversed by means of the buzzing noise produced by their wings. In fact, they do, but the whole truth proved far more complex: Bees are masters of the elegant language of dance.

Although its movements at first appear random and erratic, a forager bee reports highly specific information about a newfound food source upon returning to its hive. Its dance tells the location, type, and even the quality of the find.

If the food source is close to the hive, for example, the forager will perform a so-called round dance, during which it flies in a tight, circular pattern. The bees zero in on the food by following traces of scent provided by the forager.

When the food source is more distant, the bee must employ a more complicated "waggle dance," a tight figure-eight pattern in which the two loops are connected

A live honeybee *(above, left)* **drinks sugar water from a tube attached to a robot bee that has simulated one step of the complex bee language.**

by a straight run whose angle and length tell the direction and distance of the food. The vigor of the dance and bursts of sound convey information about the quantity and quality of the food.

Although the patterns of the dances appear to be universal, the language does have regional dialects that distinguish bees from different parts of the world. For example, an Austrian bee's waggle dance might signal its hive mates that a food source lies fifty yards away, while the same dance performed in Egypt would indicate a much closer supply. □

ACKNOWLEDGMENTS

The editors wish to thank these individuals and institutions for their valuable assistance:

Richard Alexander, Department of Zoology, University of Michigan, Ann Arbor; Bob Anderson, Helena, Montana; Bruce Beehler, National Museum of Natural History, Smithsonian Institution, Washington, D.C.; M. Boquillon, Paris; Gerald Borgia, Department of Zoology, University of Maryland, College Park; Howard W. Braham, National Marine Mammal Laboratory, National Marine Fisheries Service, National Oceanic and Atmospheric Administration, Seattle, Washington; Thomas G. Cochlin, Pacific Fleet Public Affairs, Pearl Harbor, Hawaii; T. H. Friend, Department of Animal Science, Texas A&M University, College Station; Richard J. Grojean, Weymouth, Massachusetts; Guy Hodge, Humane Society of the United States, Washington, D.C.; Humane Society of the United States, Washington, D.C.; Herman Krebs, National Zoological Park, Smithsonian Institution, Washington, D.C.; Randall Lockwood, Humane Society of the United States, Washington, D.C.; Somine Loubser, The Kaffrarian Museum, King Williams Town, South Africa; Guy-J. Mansencal, Réunion des Amateurs de Chiens Pyrénées, Tarbes, France; Madeleine Martinet, Tannay, France; Douglas Mock, Department of Zoology, University of Oklahoma, Norman; Laurence Mondet, Le Havre, France; Peter Mühling, Tiergarten, Nuremberg, Germany; David A. Nickle, National Museum of Natural History, Smithsonian Institution, Washington, D.C.; Craig Packer, Department of Ecology, Evolution and Behavior, University of Minnesota, Minneapolis; Alain Pecoult, Trésoriér, Saint Claude, France; G. Pouchin, Paris; Glenn D. Prestwich, State University of New York, Stony Brook; Peter C. H. Pritchard, Florida Audubon Society, Maitland, Florida; Sue Savage Rumbaugh, Yerkes Primate Research Center, Emory University and Georgia State University, Atlanta; Hashiro Shimuki, Beltsville Agricultural Research Center, Beltsville, Maryland; Marla Spivak, Center for Insect Science, University of Arizona, Tucson; Philip Stander, Cornell University, Ithaca, New York; Evelyne Stawicki, Service de Presse, Société Protectrice des Animaux, Paris; Pierre Swanepoel, The Kaffrarian Museum, King Williams Town, South Africa; Allen Sylvester, Agricultural Resource Lab, Baton Rouge, Louisiana; Gaynel Wald, San Juan Capistrano Mission, San Juan Capistrano, California; Penny Walker, Woodbridge, Virginia; Wayne O. Whitney, Sugarland, Texas; Kenneth R. Whitten, Alaska Department of Fish and Game, Division of Wildlife Conservation, Fairbanks, Alaska; Cathy Yarbrough, Yerkes Primate Research Center, Emory University and Georgia State University, Atlanta.

PICTURE CREDITS

Free Illustrations of Mammals, Birds, Fish, Insects, etc., selected by Jim Harter, © 1979 Dover Publications. **53:** Mark Newman/Earth Images, Washington, D.C., background, R. Fukuhara/Westlight, Los Angeles. **54:** Doug Allan/Oxford Scientific Films, Long Hanborough. **55:** Bruce M. Beehler. **56:** Yves Kerban/Jacana, Paris. **57:** Patti Murray/Animals Animals, New York. **58:** Kim Taylor/Bruce Coleman Ltd., London—Charlie Ott/Bruce Coleman Ltd., London. **59:** K. G. Preston-Mafham/Permaphotos Wildlife, Alcester, Warwick—Edward S. Ross. **60:** Roger Hoskings/NHPA, Ardingley, Sussex. **61:** Hans Reinhard/Bruce Coleman Inc., New York; H. L. Fox/Oxford Scientific Films, Long Hanborough. **62:** Jen and Des Bartlett/Bruce Coleman Ltd., London. **63:** Tom McHugh/Photo Researchers, Inc., New York. **64:** Toby Talbot—Joseph T. Collins/Photo Researchers, Inc., New York. **65:** Jack D. Swenson/Tom Stack & Associates, Colorado Springs. **66:** © Eric Hosking/Bruce Coleman Inc., New York. **67:** Dennis Green/Bruce Coleman Ltd., London. **68:** Bianca Lavies—M. J. Tyler/NHPA, Ardingley, Sussex. **69:** Douglas Mock. **70:** Otorohanga Zoological Society, Inc., New Zealand. **71:** Jean-Michel Labat/Jacana, Paris—Wild Nature/A.N.T. Photo Library. **72:** ANP-Photo, Amsterdam. **73:** © Mike Greer/Chicago Zoological Society, Brookfield Zoo—Winner/Jacana, Paris. **74:** Hans Pfletschinger/Peter Arnold, Inc., New York. **75:** Edward S. Ross. **76:** © Craig Packer 1982. **77:** Barbara Smuts/Anthro Photo, Cambridge, Massachusetts. **78:** Rod Allin/Tom Stack & Associates, Colorado Springs. **79:** Vernon Merritt for *LIFE*, background, L. Lee/Westlight, Los Angeles. **80:** George D. Lepp/Comstock, New York. **81:** Jeanne A. Mortimer. **82:** C. Allan Morgan. **83:** David C. Fritts/Animals Animals, New York. **84, 85:** Art by Time-Life Books Inc. **86:** Dwight R. Kuhn. **87:** Graham Pizzey. **88:** M. P. Kahl/Bruce Coleman Inc., New York—M. P. Kahl/DRK Photo, Sedona, Arizona. **89:** Charles Varon/Stockphoto Inc., New York. **90, 91:** © Jeff Lepore 1989. **92, 93:** C. H. Greenewalt/VIREO, Philadelphia; Vernon Merritt for *LIFE*. **94:** Tom Bledsoe/Photo Researchers, Inc., New York. **95:** © 1989 Robert A. Tyrrell—B. Schorre/VIREO, Philadelphia. **96:** Steve C. Wilson/Entheos, Seabeck, Washington—Geoff Moon. **97:** Tatarsky/Design Photographers International, Inc., New York. **98:** Larry Sherer, copied from National Archives no. 80-G-334455. **99:** Neal J. Menschel. **100, 101:** George McCarthy/Bruce Coleman Ltd., London; Abbie Zeltzer. **102:** G. Tortoli/F.A.O., Rome. **103:** S. Johnson. **104:** Fred Friberg, NN Samfoto. **105:** Photo A.F.P., Paris. **106:** Sonja Zytkow, Johannesburg, South Africa. **107:** Fil Hunter—B. Lyon/Valan Photos, St. Lambert, Canada. **108:** Grant Heilman Photography, Inc., Lititz, Pennsylvania. **109:** Enrico Ferorelli. **110, 111:** Ronald H. Cohn, The Gorilla Foundation, Woodside, California. **112:** Erwin and Peggy Bauer. **113:** Masao Kawai/Orion: Bruce Coleman Ltd., London. **114:** S. Washburn/Anthro Photo, Cambridge, Massachusetts. **115:** I. DeVore/Anthro Photo, Cambridge, Massachusetts. **116:** Yerkes Primate Research Center, Atlanta. **117:** W. McQuire/Anthro Photo, Cambridge, Massachusetts. **118, 119:** Jim Tuten/Animals Animals, New York. **120:** Philip Stander. **121:** From *Animals: 1419 Copyright-Free Illustrations of Mammals, Birds, Fish, Insects, etc.*, selected by Jim Harter, © 1979 Dover Publications. **122:** From *Animals: 1419 Copyright-Free Illustrations of Mammals, Birds, Fish, Insects, etc.*, selected by Jim Harter, © 1979 Dover Publications; Steve Rannels/Grant Heilman Photography, Inc., Lititz, Pennsylvania; Patti Murray/Animals Animals, New York. **123:** Runk/Schoenberger/Grant Heilman Photography, Inc., Lititz, Pennsylvania—from *Animals: 1419 Copyright-Free Illustrations of Mammals, Birds, Fish, Insects, etc.*, selected by Jim Harter, © 1979 Dover Publications. **124:** Erwin and Peggy Bauer. **125:** Mary Evans Picture Library, London. **126:** E. R. Degginger, FPSA Color-Pic, Inc., Convent Station. **127:** Roger Wilmburst/Bruce Coleman Ltd., London. **128:** E. R. Degginger, FPSA Color-Pic, Inc., Convent Station. **129:** John Cancalosi. **130:** P. A. Hinchcliffe/Bruce Coleman Ltd., London. **131:** © 1990 Mark W. Moffett.

BIBLIOGRAPHY

Books

Allen, Thomas B., Karen Jensen, and Philip Kopper. *Earth's Amazing Animals*. Vienna, Va.: National Wildlife Federation, 1983.

Amon, Aline. *Orangutan: Endangered Ape*. New York: Atheneum, 1977.

Anderson, E. W. *Animals as Navigators*. New York: Van Nostrand Reinhold, 1983.

Animal Architects. Washington, D.C.: National Geographic Society, 1984.

The Audubon Society Encyclopedia of Animal Life. New York: Crown, 1987.

Baker, Robin (Ed.). *The Mystery of Migration*. New York: Viking Press, 1981.

Barber, Carolyn. *Animals at War*. New York: Harper & Row, 1971.

Berrill, Jacquelyn. *Wonders of How Animals Learn*. New York: Dodd, Mead, 1979.

The Big Book of Amazing Animal Behavior. New York: Grosset & Dunlap, 1987.

Birds: Owls, Parrots & Waders (All the World's Animals series). New York: Torstar Books, 1985.

Blond, Georges. *The Great Migrations*. New York: Macmillan, 1960.

Breland, Osmond P. *Animal Life and Lore* (rev. ed.). New York: Harper & Row, 1972.

Bright, Michael:
Animal Language. Ithaca, N.Y.: Cornell University Press, 1984.
The Living World. New York: St. Martin's Press, 1987.

Burton, Maurice, and Robert Burton. *Inside the Animal World*. New York: Quadrangle/New York Times, 1977.

Burton, Robert:
Bird Behavior. New York: Alfred A. Knopf, 1985.
Eggs: Nature's Perfect Package. New York: Facts On File Publications, 1987.

The Mating Game. Oxford: Elsevier & Phaidon, 1976.

Nature's Night Life. Poole, Dorset: Blandford Press, 1982.

Caras, Roger. *The Endless Migrations.* New York: E. P. Dutton, Truman Talley Books, 1985.

Chapman, R. F. *A Biology of Locusts.* London: Edward Arnold, 1976.

Corliss, William R. *Incredible Life: A Handbook of Biological Mysteries.* Glen Arm, Md.: Sourcebook Project, 1981.

Cruickshank, Allan D., and Helen G. Cruickshank. *1001 Questions Answered about Birds.* New York: Dodd, Mead, 1958.

Davis, Flora. *Eloquent Animals: A Study in Animal Communication.* New York: Coward, McCann & Geoghegan, 1978.

Domico, Terry. *Bears of the World.* New York: Facts On File Publications, 1988.

Dossenbach, Hans D. *The Family Life of Birds.* New York: McGraw-Hill, 1971.

Douglas-Hamilton, Iain, and Oria Douglas-Hamilton. *Among the Elephants.* New York: Bantam Books, 1976.

Downer, John. *Supersense: Perception in the Animal World.* New York: Henry Holt, 1988.

Eimerl, Sarel, Irven DeVore, and the Editors of Time-Life Books. *The Primates* (Life Nature Library series). New York: Time-Life Books, 1965.

Embery, Joan, and Ed Lucaire. *Joan Embery's Collection of Amazing Animal Facts.* New York: Dell, 1983.

The Encyclopedia of Animal Ecology. New York: Facts On File Publications, 1987.

Evans, Peter. *Ourselves and Other Animals.* New York: Pantheon Books, 1987.

Ferry, Georgina (Ed.). *The Understanding of Animals.* Oxford: Basil Blackwell, 1984.

Fogden, Michael, and Patricia Fogden. *Animals and Their Colors.* New York: Crown, 1974.

Frings, Hubert, and Mable Frings. *Animal Communications* (2nd rev. ed.). Norman: University of Oklahoma Press, 1977.

Frisch, Karl von, and Otto von Frisch. *Animal Architecture.* Translated by Lisbeth Gombrich. New York: Harcourt Brace Jovanovich, 1974.

George, Jean Craighead. *Beastly Inventions: A Surprising Investigation into How Smart Animals Really Are.* New York: David McKay, 1970.

Greenhall, Arthur M., and Uwe Schmidt (Eds.). *Natural History of Vampire Bats.* Boca Raton, Fla.: CRC Press, 1988.

Grzimek, Bernhard (Ed.). *Grzimek's Animal Life Encyclopedia.* New York: Van Nostrand Reinhold, 1972.

Grzimek's Encyclopedia of Mammals. New York: McGraw-Hill, 1989.

Hahn, Emily. *Look Who's Talking!* New York: Thomas Y. Crowell, 1978.

Hansell, Michael H. *Animal Architecture and Building Behaviour.* New York: Longman, 1984.

Harrar, George, and Linda Harrar. *Signs of the Apes, Songs of the Whales.* New York: Simon & Schuster, 1989.

Hölldobler, Bert, and Edward O. Wilson. *The Ants.* Cambridge, Mass.: Harvard University Press, Belknap Press, 1990.

Hosking, Eric, David Hosking, and Jim Flegg. *Eric Hosking's Birds of Prey of the World.* Lexington, Mass.: Stephen Greene Press, 1987.

Jenkins, Alan C. *Mysteries of Nature.* New York: Facts On File Publications, 1984.

Kaufmann, John. *Insect Travelers.* New York: William Morrow, 1972.

Laycock, George. *The Story of Animal Migration: Wild Travelers.* New York: Four Winds Press, 1974.

Lockley, R. M. *Animal Navigation.* New York: Hart, 1967.

McClung, Robert M. *Mysteries of Migration.* Champaign, Ill.: Garrard, 1983.

McNamee, Thomas. *The Grizzly Bear.* New York: Alfred A. Knopf, 1984.

Maneaters and Marmosets. New York: Hearst Books, 1976.

The Marshall Cavendish International Wildlife Encyclopedia. New York: Marshall Cavendish, 1989.

Marvels and Mysteries of Our Animal World. Pleasantville, N.Y.: Reader's Digest, 1964.

The Marvels of Animal Behavior. Washington, D.C.: National Geographic Society, 1972.

Milne, Lorus, Margery Milne, and Franklin Russell. *The Secret Life of Animals.* New York: E. P. Dutton, 1975.

Moss, Cynthia. *Elephant Memories: Thirteen Years in the Life of an Elephant Family.* New York: William Morrow, 1988.

O'Toole, Christopher (Ed.). *The Encyclopedia of Insects.* New York: Facts On File Publications, 1986.

Page, Jake, and Eugene S. Morton. *Lords of the Air: The Smithsonian Book of Birds.* Washington, D.C.: Smithsonian Books, 1989.

Parfit, Michael. *South Light: A Journey to the Last Continent.* New York: Macmillan, 1985.

Parks, Peter. *Underwater Life* (The World You Never See series). Chicago: Rand McNally, 1976.

Pasquier, Roger F. *Watching Birds: An Introduction to Ornithology.* Boston: Houghton Mifflin, 1977.

Preston-Mafham, Rod, and Ken Preston-Mafham. *Spiders of the World.* New York: Facts On File Publications, 1984.

Reader's Digest Book of Facts. Pleasantville, N.Y.: Reader's Digest, 1987.

Reader's Digest Illustrated Guides to Southern Africa: Southern and Eastern Cape. Cape Town, South Africa: Reader's Digest Association South Africa, 1983.

Ricard, Matthieu. *The Mystery of Animal Migration.* New York: Hill & Wang, 1969.

Ridley, Mark. *Animal Behaviour: A Concise Introduction.* Oxford: Blackwell Scientific Publications, 1986.

Robinson, David. *Living Wild.* Vienna, Va.: National Wildlife Federation, 1980.

Rowland-Entwistle, Theodore. *Insect Life* (The World You Never See series). Chicago: Rand McNally, 1976.

Sanderson, Ivan T. *Living Mammals of the World.* Garden City, N.Y.: Doubleday, 1967.

Schauenberg, Paul. *Animal Communication* (Animal Behaviour series). Translated by R. D. Martin, F. I. Biol, and A.-E. Martin. London: Burke Books, 1981.

Scheffer, Victor B. *Spires of Form: Glimpses of Evolution.* Seattle: University of Washington Press, 1983.

Settel, Joanne, and Nancy Baggett. *How Do Ants Know When You're Having a Picnic?* New York: Atheneum, 1986.

Shreeve, James. *Nature: The Other Earthlings.* New York: Educational Broadcasting, 1987.

Siegel, Ronald K. *Intoxication: Life in Pursuit of Artificial Paradise.* New York: E. P. Dutton, 1989.

Simon, Hilda. *The Courtship of Birds.* New York: Dodd, Mead, 1977.

Sinclair, Sandra. *How Animals See: Other Visions of Our World.* New York: Facts On File Publications, 1985.

Slater, Peter J. B. (Ed.). *The Encyclopedia of Animal Behavior.* New York: Facts On File Publications, 1987.

Sparks, John. *Bird Behavior.* New York: Grosset

& Dunlap, 1970.

Sparks, John, and Tony Soper. *Penguins.* New York: Facts On File Publications, 1987.

Stonehouse, Bernard. *Young Animals: The Search for Independent Life.* New York: Viking Press, 1974.

Street, Philip. *Animal Migration and Navigation.* New York: Charles Scribner's Sons, 1976.

Terres, John K. *The Audubon Society Encyclopedia of North American Birds.* New York: Alfred A. Knopf, 1980.

Tinbergen, Niko, and the Editors of Time-Life Books. *Animal Behavior* (Life Nature Library series). Alexandria, Va.: Time-Life Books, 1980.

Topoff, Howard (Ed.). *The Natural History Reader in Animal Behavior.* New York: Columbia University Press, 1987.

Trivers, Robert. *Social Evolution.* Menlo Park, Calif.: Benjamin/Cummings, 1985.

Uvarov, Boris. *Grasshoppers and Locusts: A Handbook of General Acridology* (Vol. 2 of *Behaviour, Ecology, Biogeography, Population Dynamics).* London: Centre For Overseas Pest Research, 1977.

Waal, Frans de. *Peacemaking among Primates.* Cambridge, Mass.: Harvard University Press, 1989.

Wakefield, Pat A., and Larry Carrara. *A Moose for Jessica.* New York: E. P. Dutton, 1987.

Walters, Mark Jerome. *The Dance of Life: Courtship in the Animal Kingdom.* New York: William Morrow, Arbor House, 1988.

Wilson, Edward O. *Sociobiology: The New Synthesis.* Cambridge, Mass.: Harvard University Press, Belknap Press, 1975.

Wood, Gerald L. *Animal Facts and Feats* (rev. ed.). New York: Sterling, 1977.

Periodicals

Abrahamson, David. "Do Animals Think?" *National Wildlife,* August September 1983.

Aldhous, Peter. "Taking Notes When Birds Call the Tune." *BBC Wildlife,* October 1988.

Allen, Stevan. "On a Flight to Stardom." *St. Petersburg Times,* October 16, 1989.

Allman, William F. "Not Just Blowing Their Own Horns." *U.S. News & World Report,* May 9, 1988.

Alper, Joseph. "The Big Sting." *Health,* April 1989.

Anderson, Ian. "Sixth Sense Is the Platypus's Secret." *New Scientist,* May 12, 1988.

"The Astonishing Armadillo." *National Geographic,* June 1982.

Bachus, Richard C. "Airport Owls." *Christian Science Monitor,* April 19, 1988.

"Bait-and-Capture by an Insect in Disguise." *Science News,* 1982, Vol. 122, p. 379.

Bartlett, Des, and Jen Bartlett. "Beavers: Master Mechanics of Pond and Stream." *National Geographic,* May 1974.

"Birds That Walk on Water." *National Geographic,* May 1982.

Bower, Bruce. "Kanzi Extends His Speech Reach . . ." *Science News,* August 27, 1988.

Branigan, William. " 'Killer Bees' Appear to Be Beating the Bee Killers." *Washington Post,* November 20, 1988.

Brody, Jane E. "Dread Locusts: An Appreciation, Almost." *New York Times,* May 17, 1988.

Cater, Bill. "Just Cats." *BBC Wildlife,* October 1987.

Ciesla, Bill. "Butterfly Trees." *American Forests,* January-February 1988.

Collins, Peter. "Locusts: Vigilance Still Needed." *Nature,* May 1978.

Coonrod, Elizabeth. "Albuquerque Beset by Unexpected Bear Invasion." *Christian Science Monitor,* September 14, 1989.

Cowley, Geoffrey:
"High Times in the Wild Kingdom." *Newsweek,* January 1, 1990.
"The Wisdom of Animals." *Newsweek,* May 23, 1988.

Dunn, Euan. "Bee Raises Voice to Quell Unrest." *BBC Wildlife,* August 1988.

Eaton, Randall L. "A Possible Case of Mimicry in Larger Mammals." *Evolution,* 1976, Vol. 30, pp. 853-855.

Eckholm, Erik. "Pygmy Chimp Readily Learns Language Skill." *New York Times,* June 24, 1985.

Fisher, Allan C., Jr. "Mysteries of Bird Migration." *National Geographic,* August 1979.

Gilbert, Susan. "Wildlife on Main Street." *Science Digest,* June 1986.

Gorney, Cynthia. "The Gorilla Speaks." *Washington Post,* January 31, 1985.

Gould, Stephen Jay. "This View of Life." *Natural History,* November 1986.

Grover, Wayne. "Lost Dog's Journey of Love: 2,000 Miles on Foot to Find Her Master." *Sunday Express,* July 22, 1979.

Hansell, Michael H. "Wasp Papier-Mâché." *Natural History,* August 1989.

Hatch, R. C. "Effect of Drugs on Catnip *(Nepeta Cataria):* Induced Pleasure Behavior in Cats."

American Journal of Veterinary Research, January 1972.

"I'll Think about That Tomorrow." *Discover,* February 1990.

Jaeger, Edmund C. "Poorwill Sleeps Away the Winter." *National Geographic Magazine,* February 1953.

Kaufmann, John H. "The Wood Turtle Stomp." *Natural History,* August 1989.

Kenyon, Jean. "Northern Hemisphere Albatrosses." *Sea Frontiers,* November-December 1989.

Kruuk, Hans. "Hyenas: The Hunters Nobody Knows." *National Geographic,* July 1968.

Lessem, Don. "Here Come the Killer Bees." *International Wildlife,* May-June 1987.

Lewin, Roger. "New Look at Turtle Migration Mystery." *Research News,* February 24, 1989.

"La Longue Route du Chat Gaston." *Var Matin,* August 11, 1985.

Lucas, Jeremy. "Flight of the Arctic Tern." *Reader's Digest Canadian,* June 1984.

Lynch, Wayne. "Incredible Flights: Where, Why, How, and How Far Our Birds Migrate." *Canadian Geographic,* October-November 1985.

Marsh, Barbara. "What Is Whiskered and Ugly and Has Little Squinty Eyes?" *Wall Street Journal,* April 19, 1990.

Mock, Douglas W. "Knockouts in the Nest." *Natural History,* May 1985.

Moffett, Mark W. "Dance of the Electronic Bee." *National Geographic,* January 1990.

Molyneux, Russell J., and Lynn F. James. "Loco Intoxication: Indolizidine Alkaloids of Spotted Locoweed *(Astragalus Lentiginosus)."* *Science,* April 1982.

Mydans, Seth. "Bad Day for Capistrano? Here Come the Swallows." *New York Times,* March 19, 1990.

Myers, J. P. "Sex and Gluttony on Delaware Bay." *Natural History,* May 1986.

"Nice Nitpickers." *Discover,* November 1989.

Novacek, Michael J. "Navigators of the Night." *Natural History,* October 1988.

Palen, Gary F., and Graham V. Goddard. "Catnip and Oestrous Behaviour in the Cat." *Animal Behaviour,* April-July 1966.

Payne, Katharine. "Elephant Talk." *National Geographic,* August 1989.

Polunin, Ivan. "Who Says Fish Can't Climb Trees?" *National Geographic,* August 1983.

Pusey, Anne, and Craig Packer. "Once and Future

Kings." *Natural History,* August 1983.

Raloff, J. "Do Sea Turtles Smell the Way Home?" *Science News,* April 14, 1984.

Redmond, Ian. "The Sounds of Silence Revealed." *BBC Wildlife,* September 1988.

Roberts, Leslie. "Insights into the Animal Mind." *Bio Science,* June 1983.

Siegel, Ronald K. "Animal Intoxication: A Study in the Natural and Nonabusive Use of Drugs." *Omni,* March 1986.

Smuts, Barbara. "What Are Friends For?" *Natural History,* February 1987.

Sullivan, Walter. "Genetic Analysis Upsets Theory on Turtle Trek." *New York Times,* March 14, 1989.

"A Tall Story." *Economist,* November 21, 1987.

Underwood, Nora, and Andrea Dabrowski. "Killers on a Rampage." *Macleans,* October 3, 1988.

Vessels, Jane. "Koko's Kitten." *National Geographic,* January 1985.

Walcott, Charles. "Show Me the Way You Go Home." *Natural History,* November 1989.

Weiss, Rick. "New Dancer in the Hive." *Science News,* October 28, 1989.

Wilkinson, Gerald S. "Food Sharing in Vampire Bats." *Scientific American,* February 1990.

Yarbrough, Cathy. "GSU's Language Research Center: Language and Communication." *Research,* Fall 1985.

INDEX

Numerals in italics indicate an illustration of the subject mentioned.

ards, *38, 43;* of llamas, 38; of opossums, *39;* of owls, 42; of petrels, 38; of ratels, 42; of shrews, 42; of snakes, *38,* 42; of termites, 35; of tortoises, 38; of worms, 38. *See also* Disguises
Desert fennec foxes. *See* Foxes
Diet. *See* Appetite
Disguises: of ant lions, 46; of beetles, 45; of bugs, *45;* of butterflies, 45; of chameleons, 21; of ermines, *44;* of flies, 45; of foxes, 44; of hares, 44; of hawks, 44; of Indian sticks, 45; of katydids, 45; of lizards, 21; of mantises, 46; of moths, 45; of owls, 44; of planthoppers, 45; of ptarmigans, *44;* of snakes, *46;* of treehoppers, *45;* of turtles, 46; of weasels, 44; of weevils, 45; of worms, 45; of zebras, *47. See also* Defenses
Diving: of gannets, 26; of loons, *26;* of penguins, *9;* of seals, 10. *See also* Swimming
Dogs: and armadillos, 41; food sharing of, 126; and grouse, 42; journeys of, 105; and killdeers, 42; and porcupines, 40; and ratels, 43
Dohong (orangutan): and tools, 112
Doves: rock doves, 103
Dragonflies: eyes of, *15*
Dragons, flying. *See* Lizards
Drinking: by giraffes, *24*
Driver ants. *See* Ants
Drunkenness: of elephants, 121
Ducks: eggs of, 128; language of, 128; mallard ducks, 128; migration of, *96*
Dunnocks: and cuckoos, 67

E

Eagles: and monkeys, 124
Ears: of bats, 16; of elephants, 18; of foxes, *18;* of jerboas, 18; of owls, 17; of rabbits, 18
Earthworms. *See* Worms
East Africa: elephants in, *120;* planthoppers in, 45; rats in, 73; wildebeest in, *97*
Eastern kingbirds. *See* Kingbirds
East Indian paradise tree snakes. *See* Snakes
Echolocation: of bats, 16; defined, 16
Eggs: of blackbirds, 128; of chickens, *58,* 128; of cuckoos, 67; of ducks, 128; of frogs, 68; of guillemots, *58;* of kiwis, *70;* of murres, 58; of turtles, *81*
Egrets: great egrets, *69;* and siblings, 69
Egypt: bees in, 131
Egyptian spiny mice. *See* Mice
Elephants: African elephants, 18; birth of, 66;

and crocodiles, 30; and Harvey Croze, 118; drunkenness of, 121; ears of, 18; in East Africa, *120;* and giraffes, 24; grief of, *118-119;* infrasounds of, 120-121; language of, 120-121; mating of, 121; and mgongo, 121; and midwifery, 66; and Katharine Payne, 120-121; in Serengeti National Park, 118; and Ronald Siegel, 121; stress of, 121; teeth of, 25; trunks of, 25; tusks of, 25; in Zimbabwe, *118-119*
Elephant seals. *See* Seals
Elk: migration of, 108
Emperor penguins. *See* Penguins
Emus: in Australia, 27; flightlessness of, 27; history of, 27
Emu War, 27
England: foxes in, 99; hedgehogs in, *100*
Ermines: disguises of, *44;* Eurasian ermines, *44*
Ethiopia: locusts in, *102*
Eucalyptus leaves: and koalas, *23*
Eurasian ermines. *See* Ermines
Europe: cuckoos in, 67
European gulls. *See* Gulls
European marsh warblers. *See* Warblers
European white storks. *See* Storks
Eyes: of bats, 15; of bees, 15; of chameleons, 15; of dragonflies, *15;* of flies, 13; of owls, *17. See also* Blindness

F

Fairy armadillos. *See* Armadillos
Falls: of cats, *19*
Farallon Islands: seals on, *65*
Fat-tailed shrews. *See* Shrews
Feathers. *See* Plumage
Fennec foxes, desert. *See* Foxes
Finches: on Galápagos Islands, 37; and tools, 37; woodpecker finches, 37
Fins: of mudskippers, 28
Fireflies: appetite of, 42; and luciferase, 15; and luciferin, 15; *Photuris* fireflies, 42
Fishers: and porcupines, 40
Fishing: by spiders, 50
Fleas: launching of, *12;* and mice, 14; and rats, 14; and resilin, 12; and Miriam Rothschild, 12
Flies: and bacteria, 13; and caribou, 83; dance fly, *59;* and DDT, 13; disguises of, 45; eyes of, 13; hangingflies, 59; houseflies, *13,* 59; reproduction of, 13; sensors of, 13; warble flies, 83; wings of, 13
Flight: of bats, *16;* of colugos, 29; of lizards, 29; of loons, 26; of snakes, 29; of squir-

rels, *29. See also* Flightlessness; Wings
Flightlessness: of emus, 27; of kiwis, 27; of ostriches, 26-27; of penguins, 27. *See also* Flight; Wings
Flower mantises, pink. *See* Mantises
Flowers: and mites, 14
Flying dragons. *See* Lizards
Flying squirrels. *See* Squirrels
Food. *See* Appetite
Food sharing: of bats, 126; of chimpanzees, 126; of dogs, 126; of hyenas, 126
Formic acid: and ants, 35
Fossey, Dian: and gorillas, 117
Fossils: of platypuses, 8
Foster parents: and cuckoos, 67
Foxes: arctic foxes, 44; desert fennec foxes, *18;* disguises of, 44; ears of, *18;* in England, 99; and grouse, 42; and killdeers, 42
Franco-Prussian War: pigeons in, 103
Freezing: of beetles, 9; of frogs, 9; of squirrels, 31
Friendships: between baboons, 77
Frogs: African clawed frogs, 36; Australian gastric-brooding frogs, *68;* and beetles, 39; freezing of, 9; and glucose, 9; North American wood frogs, 9; and shrews, *43;* and snakes, *46;* thawing of, 9
Fundy, Bay of: terns in, *86*
Fungus: and ants, 34
Fur seals. *See* Seals

G

Galápagos Islands: finches on, 37
Gannets: diving of, 26
Gaston (cat), *105. See also* Cats
Gastric-brooding frogs. *See* Frogs
Gazelles: and baboons, *114*
Gecko lizards. *See* Lizards
Geese: migration of, *96*
Gerbils: and exercise wheels, *107;* migration of, 107
Germany: pigeons in, 103
Giant flying squirrels. *See* Squirrels
Giant petrels. *See* Petrels
Gills: of mudskippers, 28; of perch, 28
Giraffes: birth of, 79; drinking water by, *24;* and elephants, 24; hearts of, 25
Glucose: defined, 9; and frogs, 9
Glycerol: and beetles, 9; defined, 9
Goats: appetite of, 36; and James Cook, 36; in Morocco, *36;* in New Zealand, 36; in trees, *36*
Goodall, Jane: and chimpanzees, 116, 117
Gooney birds. *See* Albatrosses

Llamas: defenses of, 38
Locoweed: and horses, *123*
Locusts: in Africa, 102; in Beartooth Mountains, 106; in Ethiopia, *102;* in Grasshopper Glacier, 106; in India, 102; in Middle East, 102; migration of, *102,* 106; in Montana, 106
Lodges: of beavers, *78*
Loons: diving of, *26;* flight of, 26
Lorenz, Konrad: and imprinting, 61
Luciferase: and fireflies, 15
Luciferin: and fireflies, 15
Lugworms. *See* Worms

M

Macaques: on Koshima Island, *113*
Macrotermes. See Termites
Madagascar: tenrecs in, 72
Maine: moose in, 99
Malaysian Basicerotine ants. *See* Ants
Mallard ducks. *See* Ducks
Mallee fowls: in Australia, 62; nests of, *62*
Mantises: disguises of, 46; pink flower mantises, 46
Marsh warblers, European. *See* Warblers
Marsupials. *See* Kangaroos; Koalas; Opossums
Masked weaverbirds. *See* Weaverbirds
Massachusetts: owls in, 99
Mating: of baboons, 77, *115;* of beavers, 78; of crabs, 91; of crab spiders, 61; of dance flies, 59; of elephants, 121; of hanging-flies, 59; of jackdaws, *61;* of penguins, 54; of seals, 65; of weaverbirds, 59. *See also* Courtship; Reproduction
Maypoles: of bowerbirds, 57
Mealybugs: and ants, 34; sap-feeding mealybugs, 34
Megaponera ants. *See* Ants
Mexico: ants in, 34; bees in, 101; butterflies in, *80;* seals in, 65; turtles in, *82*
Mgongo: and elephants, 121
Mice: and beetles, 14; birth of, 66; Egyptian spiny mice, 66; and exercise wheels, 107; and fleas, 14; harvest mice, *60;* migration of, 107; and mites, 14; nests of, *60*
Middle East: locusts in, 102; snakes in, 46
Midway Island: albatrosses on, *98*
Migration: of albatrosses, 84, 88, 98; of birds, *map* 84-85; of bobolinks, 92; of bugs, 108; of butterflies, *80;* of caribou, *83;* of chickadees, 108; of crabs, *90-91;* of cuckoos, 96; of deer, *108;* of ducks, *96;* of elk, 108; of geese, *96;* of gerbils, 107; of hamsters, 107; of hippopotamuses, 106; of

hummingbirds, 95; of kingbirds, 95; of lemmings, *104;* of locusts, *102,* 106; of mice, 107; of moose, 108; of owls, 99; of penguins, 27; of seals, 94; of shearwaters, 84, 87; of songbirds, 92-93; of storks, 85, 88; of swallows, *89;* of swans, *92-93;* of terns, 85, 86-87; of turtles, 81, *82;* of warblers, 92, 93; of wildebeest, *97*
Milk: of kangaroos, 70-71
Mimus polyglottos. See Mockingbirds
Mission Viejo: and swallows, 89
Mites: and flowers, 14; and hummingbirds, 14; and mice, 14; and rats, 14; and spiders, *14*
Mockingbirds: language of, 129
Mole rats, naked, 73
Moles: blindness of, 20; noses of, *20;* star-nosed moles, *20;* and worms, 20
Monarch butterflies. *See* Butterflies
Mondet, Laurence, *105*
Monkeys: and eagles, 124; language of, 124; and leopards, 124; and snakes, 124; vervet monkeys, 124
Montana: locusts in, 106
Moose: in Alaska, 99; antlers of, 64; courtship of, 64; in Maine, 99; migration of, 108; in New Hampshire, 99; in New York State, 99. *See also* Joshua (moose)
Morocco: goats in, *36*
Mosquitoes: and caribou, 83
Moths: and bats, *16;* Burmese moths, 45; disguises of, 45; and spiders, 50; *Stenoma algidella* moths, 45
Mountain chickadees. *See* Chickadees
Mountain lions: and wolverines, 43
Mudskippers: courtship of, *28;* fins of, 28; gills of, 28; and insects, 28; tails of, 28; tree climbing of, *28;* and worms, 28
Mule deer. *See* Deer
Murres: eggs of, 58
Mushrooms: *Amanita muscaria,* 122; and reindeer, 122
Myna birds: and Calvin Coolidge, 129; language of, 129
Myths: about animals, 21

N

National Zoo, 129
Navajo: and poorwills, 32
Ndutu Lake: wildebeest in, *97*
Nepeta cataria. See Catnip
Nepetalactone: and cats, 122
Nests: of ants, *34;* of bowerbirds, 57; of hornbills, 62-63; of mallee fowls, *62;* of mice,

60; of swallows, 89; of swiftlets, *66;* of wasps, *74;* of weaverbirds, 59
New Guinea: birds of paradise in, 55; bowerbirds in, 57; weevils in, 45
New Hampshire: moose in, 99
New Jersey: crabs in, *90-91*
New Mexico: bears in, 99
New York State: moose in, 99; pigeons in, 105
New Zealand: dance flies in, 59; goats in, 36; kiwis in, 27, 36, 70
Nick (dog): journey of, 105. *See also* Dogs
North Africa: snakes in, 46
North American black bears. *See* Bears
North American flying squirrels. *See* Squirrels
North American jack rabbits. *See* Rabbits
North American wood frogs. *See* Frogs
North American wood turtles. *See* Turtles
North American zone-tailed hawks. *See* Hawks
Norway: lemmings in, *104*
Noses: of camels, *7, 22;* of moles, *20*
Nuttall's poorwills. *See* Poorwills

O

Offspring, raising of: and bears, 68; and penguins, *54-55*
Olympic Games, 103
Opossums: defenses of, *39;* Virginia opossums, *39*
Orangutans: and tools, *112*
Osten, Wilhelm von: and Clever Hans, *125*
Ostriches: and antelopes, 52; flightlessness of, 26-27; history of, 27; myths about, 21; running of, *26-27;* and zebras, 52
Otters: sea otters, *37;* and tools, 37
Owls: and airports, *99;* and badgers, 42; brains of, 17; burrowing owls, 42; and cats, 42; and coyotes, 42; defenses of, 42; disguises of, 44; ears of, 17; eyes of, *17;* great horned owls, *17;* in Massachusetts, 99; migration of, 99; and porcupines, 40; and Norman Smith, *99;* and snakes, 42; snowy owls, 44, *99*
Oxpeckers: and buffalo, *51*

P

Pancake tortoises, African. *See* Tortoises
Panthers: black panthers, *21*
Paper: and wasps, *74*
Paradise tree snakes, East Indian. *See* Snakes
Parrots, *129;* language of, 129
Patterson, Francine: and Koko, 110-111
Payne, Katharine: and elephants, 120-121
Peacocks: courtship of, 61

Time-Life Books Inc.
is a wholly owned subsidiary of
THE TIME INC. BOOK COMPANY

President and Chief Executive Officer:
Kelso F. Sutton
President, Time Inc. Books Direct:
Christopher T. Linen

TIME-LIFE BOOKS INC.

EDITOR: George Constable
Director of Design: Louis Klein
Director of Editorial Resources: Phyllis K. Wise
Director of Photography and Research:
John Conrad Weiser

PRESIDENT: John M. Fahey, Jr.
Senior Vice Presidents: Robert M. DeSena,
Paul R. Stewart, Curtis G. Viebranz, Joseph J. Ward
Vice Presidents: Stephen L. Bair, Bonita L.
Boezeman, Mary P. Donohoe, Stephen L. Goldstein,
Andrew P. Kaplan, Trevor Lunn, Susan J. Maruyama,
Robert H. Smith
New Product Development: Trevor Lunn,
Donia Ann Steele
Supervisor of Quality Control: James King

PUBLISHER: Joseph J. Ward

Editorial Operations
Production: Celia Beattie
Library: Louise D. Forstall
Computer Composition: Deborah G. Tait (Manager),
Monika D. Thayer, Janet Barnes Syring,
Lillian Daniels

Library of Congress
Cataloging in Publication Data
Amazing Animals / by the editors of Time-Life
Books.
p.   cm.   (Library of curious and unusual facts).
Includes bibliographical references.
ISBN 0-8094-7695-9 (trade)
ISBN 0-8094-7696-7 (LSB)
1. Animals—Miscellanea—Popular works. [1. Ani-
mals—Miscellanea.] I. Time-Life Books. II. Series.
QL50.A43 1990
591—dc20   90-11197 CIP

LIBRARY OF CURIOUS AND UNUSUAL FACTS

SERIES DIRECTOR: Russell B. Adams, Jr.
Series Administrator: Elise Dawn Ritter-Clough
Designer: Susan K. White
Associate Editor: Sally Collins (pictures)

Editorial Staff for *Amazing Animals*
Text Editor: John R. Sullivan
Researchers: Sydney J. Baily, Regina M. Dennis
Researcher/Writer: Debra Diamond Smit
Assistant Designer: Alan Pitts
Copy Coordinators: Jarelle S. Stein (principal),
Donna Carey
Picture Coordinator: Jennifer Iker
Editorial Assistant: Terry Ann Paredes

Special Contributors: William Barnhill,
Don Oldenburg, Nancy Shute, Dan Stashower (text);
Andra H. Armstrong, Catherine B. Hackett, Eugenia
S. Scharf, Lauren V. Scharf (research); Hazel
Blumberg-McKee (index)

Correspondents: Elisabeth Kraemer-Singh (Bonn),
Christine Hinze (London), Christina Lieberman (New
York), Maria Vincenza Aloisi (Paris), Ann Natanson
(Rome).
Valuable assistance was also provided by Peter Haw-
thorne (Johannesburg); Judy Aspinall (London);
John Dunn (Melbourne); Elizabeth Brown (New
York); Dag Christensen (Oslo); Miriam Murphy, Ann
Wise (Rome); Dick Berry (Tokyo); Traudl Lessing
(Vienna).

The Consultants:
Ralph Bram, Ph.D., is a national program leader for
Medical and Veterinary Entomology and Parasitology
at the U.S. Department of Agriculture. He has published more than fifty scientific articles on various
aspects of medical and veterinary entomology, and
has participated as a consultant for the Food and
Agriculture Organization, the International Atomic
Energy Agency, the European Economic Community,
and the Pan American Health Organization.

William R. Corliss, the general consultant for the
series, is a physicist-turned-writer who has spent the
last twenty-five years compiling collections of
anomalies in the fields of geophysics, geology, ar-
chaeology, astronomy, biology, and psychology. He
has written about science and technology for NASA,
the National Science Foundation, and the Energy
Research and Development Administration (among
others). Mr. Corliss is also the author of more than
thirty books on scientific mysteries, including *Mysterious Universe, The Unfathomed Mind,* and *Handbook of Unusual Natural Phenomena.*

Irven R. DeVore, Ph.D., is chairman of the Anthropology Department at Harvard University and is a
professor in both anthropology and biology. He has
published extensively on primate behavior.

Ed Ross, Ph.D., is curator of entomology emeritus
for the California Academy of Sciences and a research associate of entomology at the University of
California. Traveling throughout the tropical world
to pursue his research, Dr. Ross helped pioneer can-
did closeup nature photography.

Marcello Truzzi, a professor of sociology at Eastern
Michigan University, is director of the Center for
Scientific Anomalies Research (CSAR) and editor of
its journal, *Zetetic Scholar.*

Charles Walcott, Ph.D., is the executive director of
the Cornell University Laboratory of Ornithology and
a professor of biology at Cornell, specializing in
neurobiology and behavior. He has published numerous articles on pigeon homing and vibration
receptors of spiders for several periodicals.

Don E. Wilson, Ph.D., is director of the Neotropical
Biodiversity Program for the Smithsonian Institution
and former chief of the Biological Survey for the
U.S. Fish and Wildlife Service at the National Museum of Natural History. He has authored more than
100 research papers on tropical biology.